HUMAN EVOLUTION, BIOLOGICAL AND CULTURAL DOMAINS

OUR ANIMAL CONDITION AND SOCIAL CONSTRUCTION

HUMAN EVOLUTION, BIOLOGICAL AND CULTURAL DOMAINS

Additional books and e-books in this series can be found on Nova's website under the Series tab.

HUMAN EVOLUTION, BIOLOGICAL AND CULTURAL DOMAINS

OUR ANIMAL CONDITION AND SOCIAL CONSTRUCTION

JORGE A. COLOMBO, MD, PHD

nova science publishers
New York

NOTICE TO THE READER

Library of Congress Cataloging-in-Publication Data

ISBN: 978-1-53615-357-6

Published by Nova Science Publishers, Inc. † New York

With love,
to my wife, Beatriz
to my parents and my brother, Emilio
to my children and grandchildren.

CONTENTS

PRELIMINARY QUOTES AND INTRODUCTORY REMARKS

"'You,' your joys and your sorrows, your memories and your ambitions, your sense of personal identity and free will, are in fact no more than the behavior of a vast assembly of nerve cells..." (Francis Crick, 1994).

"As opined by McConkey and Goodman [1997]: 'Comparative analysis of human and ape genomes is far more than an excursion into natural history at the molecular level. Until we have a detailed understanding of the genetic differences between ourselves and our closest evolutionary relatives, we cannot really know what we are'" (Kehrer-Sawatzki and Cooper, 2007).

Which and how much of our current drives – individually and as a global community – are driven by ancestral, inherited traits imprinted in our animal condition? An attempt to approximate this intriguing query is proposed here. It pertains to our identity and social constructions, and ecological interaction.

As they relate to our collective behavior, two early warnings regarding the state of the world and our species should "mark the tone":

"I do not wish to seem overdramatic, but I can only conclude from the information that is available to me as Secretary-General, that the Members of the United Nations have perhaps ten years left in which to subordinate their ancient quarrels and launch a global partnership to curb the arms race, to improve the human environment, to defuse the population explosion, and to supply the required momentum to development efforts. If such a global partnership is not forged within the next decade, then I very much fear that the problems I have mentioned will have reached such staggering proportions that they will be beyond our capacity to control" (U Thant, 1969).

"We affirm finally that any deliberate attempt to reach a rational and enduring state of equilibrium by planned measures, rather than by chance or catastrophe, must ultimately be founded on a basic change of values and goals at the individual, national, and world levels. ... The last thought we wish to offer is that man must explore himself -his goals and values- as much as the world he seeks to change. The dedication to both tasks must be unending. The crux of the matter is not only whether the human species will survive, but even more, whether it can survive without falling into a state of worthless existence" (The Executive Committee of The Club of Rome, 1972).

PREFACE

The history of our evolution over the last 10,000 years has been primarily about our cultural development and our adaptations to social contexts, and to new, natural and artificial environments. Various beliefs grown on unverifiable grounds tend to divorce our species from becoming *Natural History*. Hence, this justifies the following brief, initial comments.

As part of our biological nature, what do we have in common with other mammalian species in our physical and physiological construction? Although the answers can be found in most high school biological textbooks, it seems appropriate to briefly mention them here as they relate to future treatment in the following chapters.

1. Anatomical segments shared with other species are initially expressed in our embryological and postnatal development, followed by a process of atrophy [remains of them are found in the coccyx – a fetus with a caudal appendix or tail during embryological stages 14-22, followed by partial reabsorption –; auricular muscles attached to our ears; appendix; branchial foldings (pharyngeal)]. Comparative embryology and anatomy enlighten our true evolutive origin. If in place of developing a coccyx bone – the evolutive remains of a tail- and have a tail in its full expression, *creationists* would have to improve even more

their fantastic constructions to negate our animal origin (Colombo, 2017).

2. Our body is composed of organs, regulatory mechanisms, and systems on which our health, reproduction, well-being and survival depend on – a body that provides us with internal and external inputs and needs that compels us to satisfy them. Its construction depends on gestational and postnatal developmental periods, whose durations are characteristic of species, before achieving biological and social autonomy.
3. Our species has developed a brain where thoughts, introspective and exteroceptive actions are initiated as part of the construction of our identity. It needs oxygen and energy sources provided by a complex net of blood vessels and intracellular means; it processes stimuli and signals for our self-reference and memory, and our manifest and introspective behaviors. Its function is based on codifying and de-codifying processes, the *sine qua non*-requirements to apprehend environmental cues and to exchange information. Sensorial and perceptual processes are common to animal species and characteristic for each of them.

Through our long and hazardous evolution, several events contributed to the construction of culture (see later) that is in constant interaction to harmonize innate tendencies – "needs" resulting from our animal core – with cultural-environmental events:

1. The emergence of the neocortex with circuits whose computational complexity, organization, and neuroplasticity, allowed the development of language, sophisticated social structures, instruments and virtual constructions that generate cumulative culture. This also allowed to constantly reformulate our representations of reality and ourselves, and the possibility of developing mechanisms that made it possible to develop complex inhibitory processes – internal "vetoes" – to primary impulses and needs.

2. The postnatal extension of the period of cerebral organization (*neoteny*) made possible the social carving of the mind – socialization/domestication – and adjustments through learning of our complex social behaviors from an early age.
3. Language, which in addition to saving us much time in certain social bonding tasks, allowed us to develop complex structures of thought and collective behaviors of unprecedented sophistication in the *Natural Kingdom.*
4. Community settlements and a social structure supplied enough security – from predators and competitors – to have the time necessary to begin a great adventure of creativity and knowledge. It optimized the chances of raising an immature baby for a comparatively long period, which contributed to early brain/mind development. This allowed the fostering of knowledge and the early development and exercise of imagination and creativity. Although, these are sometimes menaced by physical, emotional, magical, or autocratic constructions in different strata of the social structure.

Figure 1. Recreation of an "iceberg" to suggest the approximate relative percentage of conscious activity (estimated as *working memory*) if this were equivalent to 12.5 percent of an iceberg. Researchers affirm that the percentage of conscious activity (working memory) could be much lower. See below in the text concerning energy consumption.

However, not all events are of a conscious dimension. Cognitive processing involves distributed neural circuits as a substrate. The most disturbing from an intellectual point of view is that much of the former appears to be at the *subconscious* level. According to some authors, what appears at the conscious level, expressed temporarily at a specific time, are the events we can manipulate as a *working memory* in our executive behavior domain and corresponding to the *explicit memory* (Dietrich, 2015). For comparison purposes, this would represent much less than the 10-15 percent that emerges from an iceberg on the surface (Dietrich, 2015) (Figure 1).

Primitive *Homo* culture was expressed in the original, ancestral communities based on standard parameters to other mammalian species: hierarchies, group priorities, survival drive, territorialism, food search, and reproductive strategies. Progressively, cultural complexity and sophistication of those basic tenets resulted in collective “moral sets” and their institutionalization. These were incorporated into the cultural heritage, as language and symbolic structures enriched communication and complex social organizations – subtle changes in brain organization mediated through processes of neuroplasticity made possible such development. It included adaptations to afferences from the physical and social environment in what configures a biosocial interaction.

Such development continued until reaching a point of cognitive development that allowed virtual constructions to be incorporated into its perception of *reality* and beliefs. Afterward, a virtual parallel dimension to the *Natural History* pervaded human cognition, foreign to its natural origin, and sometimes fed by hallucinations and behavioral disorders. Its strength merged from the construction of an imaginary dimension where supernatural beliefs offered a sense of security and expectations during a human lifetime and beyond death. It also created a moral framework to build human relationships and a place to resolve the notion of fragility, fears, sense of guilt, and deep uncertainties.

The animal origin remained occult or rejected in such para-biological constructions that occasionally reached the condition of fundamentalisms. These were able to display various worrying levels of vicious belligerence and cruelty, curtail human freedom of thought, and share natural material inequities. In attempting to "shake off" some of the restraining chains that walked with us during our *Homo* evolution, we fell prisoners to deeply entrenched emotional constructions.

Our species' knowledge, creativity, and technological developments gave support to a dominant macro species that at the same time tried to avoid attaining full consciousness of its animal origin and fundamental biological nature – a condition that affects some of our emotional expressions and social drives (territorialism, reproductive and feeding priorities, survival, prevalence). The belligerence, cruelties, social inequities, and unrelenting individual and class ambitions are the best testimony that to change our ancestral drive we must first recognize them and assume our fundamental nature. Profound cultural changes are only possible and enduring if we come to grips with our true primary condition.

To what extent does our animal condition affect our individual and social behaviors? To what extent such primary condition is considered in the construction of ideologies and social and political systems and policies? Do our pedagogic strategies include some consideration for our natural construction? Can we regenerate humanly individual and social values without considering it?

In the following chapters several issues are considered, combined with comparative evolutive considerations regarding humans and other close members of our natural clade. The aim is to provide additional views to understand better the construction of the social person and the potential risks that would depend on ancestral territorial and prevalent drives imprinted in our brains and behavioral trends.

In this attempt, it remains clear that the template of our fundamental nature did not emerge from a *tabula rasa* condition, but from endless previous evolutive events that included genetic pool exchanges, violent climate, geophysical conditions, and our early condition of prey and predator. From such a multifactorial equation appeared several *Homo* species now extinct, until the birth of an expansive, predominant species. Let us call it the *Human factor*; one that genetic paleoanthropology suggests that its true origin involves more *Homo* species than the ecumenic cultural construction of linear evolution. It can be said that several evolutionary hands rocked our *ancestral cradle*.

Hence, the issue of the human being traverses several levels of consideration in the crossroads of *nature* and *culture*. In such an interaction, the construction of identity arises. In this respect, it is interesting the comments made by Eyal Weizman (2016) in his article "*Are they human?*":

> "All that argued for the orangutan's humanity had to concede that at the very least, compared to the human, the orangutan was imperfect. In the words of American primatologist Robert Yerkes (1925, cit., by Weizman, e-flux) being "almost human", apes were perceived to be less than human. There were, of course, other groups considered by Europeans to be imperfectly human as well, like indigenous people and slaves."

To which he added:

> "Rousseau's argument (Discourse on the Origin of Inequality, 1754) for the humanity of the orangutan was not based on physical and behavioral resemblance... but rather that the ape shared a certain elasticity with humans..."

With time, the difference between human being and non-human primate suffered from a high degree of a subjective component and conceptually expressed, biological sliding into one or the other direction. Although today several hominin tribes (Figure 2) can be differentiated, during the XVIII century, concepts and hypothesis lacked the statistical demands and methodological and technical resources of today's science. Such conditions made those proposals more permeable to ideology and prejudices or trends that prevailed at the time.

In this regard, from a taxonomic point of view, it is still debatable whether the Great Apes belong to the genre *Homo*.

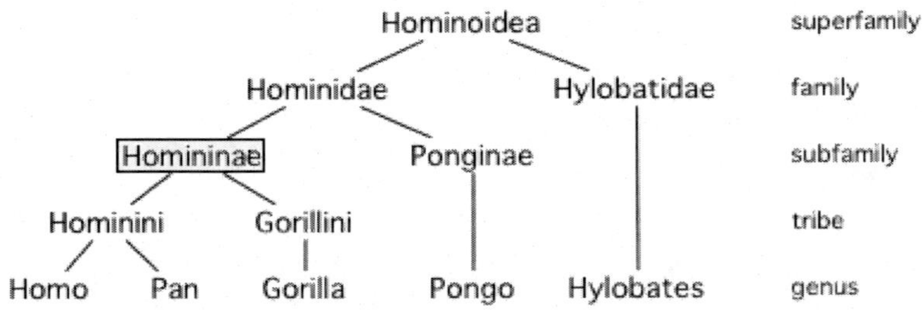

Figure 2. Taxonomic distribution (https://es.wikipedia.org/wiki/Homininae).

The evolution of our species and our civilization took place between two differing vectors: cooperation and competition. The concept of predation applied to the *Natural Kingdom* appears as anticipation to our species' cultural predatory behaviors. On these grounds it seems pertinent to include a thought from Thorstein Veblen (1899), referred to a culture's predatory phase:

> "The predatory phase of culture is attained only when the predatory attitude has become the habitual and accredited spiritual attitude for the members of the group …"

In *A Discourse on the Origin and the Foundation of the Inequality among Mankind* (1755. cf. Weizman, 2016), Jean-Jacques Rousseau delivered some concepts that appear to conserve validity:

> "I conceive two species of inequality among men; one which I call natural, or physical inequality, because it is established by nature, and consists of the difference of age, health, bodily strength, and the qualities of the mind, or of the soul; the other which may be termed moral, or political inequality, because it depends on a kind of convention, and is established, or at least authorized, by the common consent of mankind. This species of inequality consists in the different privileges, which some men enjoy, to the prejudice of others, such as that of being richer, more honored, more powerful, and even that of exacting obedience from them."

Beyond taxonomic classifications, from a sociological point of view, human prejudices feed on common unconscious sources, where some are distorted views of so-called “second-class citizens” which project on to the general social construction. A distortion that assumes behaviors as being natural or intrinsically characteristic, when they are socially conditioned. Poverty, indigence, marginalization, and slavery represent a crystallization of such prejudices supported by a financial/mercantile view of human relations.

Two additional comments may represent the views from two different times. At the beginning of the XX century, three chimpanzees would have been inseminated with semen from African persons (cf. Geus, 1997; quoted by Brüne, 2004), with the apparent intention of generating a vigorous and tamable source of free work labor. Two of the simians died due to dysentery during their transport to Europe, and the third one died in a Marseille, France zoo. On the other hand, one hundred years later, some authors claim – based on concepts of genetic anthropology – that chimpanzees belong to the same genre *Homo* as humans do. In this respect, some authors support the idea that a two million year period would be needed to define a genetic divergence of species. Should this be sustained, several fossilized *Homo* species ought to be considered as *H. sapiens* (see in Curnoe and Thorne, 2003).

Such disparity of current scientific thoughts applied to evolutive issues has been used to support social and ideological movements involved with human rights, either as segregationist or inclusive. Within this context, it ought to be mentioned an event that had an impact on social anthropology toward the end of the XIX century and the construction of political thoughts at the beginning of the XX century. Such an event was produced in part by Broca’s classification of cranial capacity in several human “races” (ethnic). It implied a clear underestimation of the relative cranial capacity of Native American and African ethnicities, compared with those belonging to Europeans or Arians. Gould (1981) (see discussion in Jensen, 1982; Carroll, 1995) severely criticized the methodologies and interpretations.

Besides giving place to prejudices and ideologic constructions, our pertinence to the animal kingdom and particularly our ape heritage ignited a rejection and a formulation of fantastic and imaginative suppositions that attempt to move the issue away from an evolutionary link.

What degree of plasticity do our behaviors have that impact the extra personal universe? Are they associated – or dependent of – an ancestral neurobiological substrate modulated by cultural history, or do they have exclusive origin on the actual socio-cultural experience and lack any evolutive inertia?

It is well known of the existence of neural circuits linked to emotional behavior. They include components associated with violence linked to social prevalence, reproductive behavior, territoriality, and access to food resources. On a group behavior bases, according to Wranghan (1999), killing coalitions among neighboring communities take place regularly in several species (wolves, chimpanzees). These are attributed – according to the imbalance of power hypothesis – to the expression of a dominant drive over neighbors. Among humans, such inter- and intragroup competence would take place on a similar basis but can be triggered by additional reasons, whether they be for survival or cultural reasons (religious, financial, ideological, expression of tribal parochialism associated to various possible slogans). The final expression of emotional behavior (whether aggression or flight) respond through a common, final neurobiological pathway.

Based on comparative behavioral analysis of intergroup belligerence in chimpanzees and hunter-gatherers, Wrangham and Glowacki (2012) suggest that the model based on a chimpanzee's behavior represents an acceptable basis for approaching an understanding of the biological and cultural evolution of armed conflicts. Bonobos (*Pan paniscus*) and gorillas express different strategies for the resolution of disputes, as will be seen later.

According to Boehm (2012), the evolutive predecessor – the ancestral *Pan* of hominids – to the *Australopithecine/Homo* lineage, the chimpanzee (*Pan troglodytes*, or common chimpanzee) and the bonobo (*Pan paniscus*, or pygmy chimpanzee) would have lived in social constructions based on hierarchical dominance. Remarkably, there is a sharp difference in conflict management exerted by chimpanzees (tendency to conflict, male predominance) and bonobos (preventive behavior, female predominance). Based on the behaviors of both species, Boehm (2012) poses that humans would keep both alternatives; a potential behavioral bipolarity with uneven prevalence distribution among individual characters and social organizations. This hypothetical behavioral duality could be linked to the configuration of the human genome and the subsequent neurobiological scaffolding. It would then be reasonable to expect that those basic tendencies could predominate in different groups and individuals, in addition to cultural conditioning. Their final behavioral phenotype would depend on socio-cultural and socio-biological conditions.

Our species' distant origin emerges from a naturally violent context that challenged individual and species survival. Intra-group and intra-species cooperation, coupled with flight and aggressive behaviors, provided relative conditions to endure. Hence, it is not completely surprising that eons later, violent behavior remains as an expression in our civilization.

From a physiological point of view, neurobiological circuits resident in limbic structures (amygdala, hypothalamus) and prefrontal cerebral cortex (Blair, 2004; Gobrogge et al., 2007; Siever, 2008), as well as hormonal influences (Rose et al., 1972; Gordon et al., 1979; Westlye et al., 2017) are involved[1]. Socially based events acting on this biological scaffolding should not be discarded, as it is the case of its influence on sexual hormonal regulation which has been shown in various species. It is interesting the individual human differences on hormonal response, as is the case of the effect of testosterone on aggressive behavior, reported by Cueva et al. (2017).

In the *Natural Kingdom*, all species are programmed to secure relative prevalence to assure essential survival components: territory where to obtain nutrients for survival and reproductive opportunities. Their behavior

---

[1] Androgenic receptors are distributed among several brain structures (amygdala, hippocampus, hypothalamus, cerebral cortex, brainstem).

– specific for each species and ecological niche – are aimed at maximizing success probability, otherwise befalls extinction and replacement.

From those premises emerged our species, imprinted through evolutionary times in our basic behavioral mechanisms. Phenotypic variations in form, physiological limits, and strategies do not surrender to those essential survival pillars. Humans have added cultural values and beliefs that tend to mask the presence of those ancestral drives. The various socio-cultural systems attempt to reformulate, sublimate and camouflage primary drives. All gregarious communities with a hierarchical organization are built upon some form of authority ranking and behavioral rules if these are violated.

Other premises entered the picture when compared with non-human animals with evolving socio-cultural constructs; vanity, fundamentalisms, consumerism, extreme forms of poverty and social marginalization, corrupted dealings with impunity, environmental abuse, disinformation, Machiavelism. Most of these are associated with the development of the social brain and material and virtual cultural constructions. Are they sophisticated and updated expressions of the ancestral menu imprinted in our basal neural configuration? Are our behavioral deviations – such as cruelty and slavery – connected to such basal conducts? When inequity and socioeconomic conditions reach certain limits, docs our old menu emerge driving our violent responses?

If we carefully observe, we can see that power, prevalence, territorial behavior and privileges also feed behaviors shared by other species, although we construct social norms and virtual agents aimed at buffering such circumstances. Primary forms of dealing with disputes involve either aggression, consensus, or submissiveness. Do they represent the spectrum of basic behaviors associated with species ancestral to *H. sapiens* and non-human primates? Gorillas and chimpanzees in a community express political actions aimed at obtaining sexual, feeding, or hierarchical privileges (De Waal, 2007).

De Waal (2005) provides a skeptical view regarding the possibility that the human species could keep evolving according to past dynamics. The author claims we have manipulated the changing environment to our advantage, thus suppressing the main handle that biology can implement as a tool to modify us through differential reproduction.

Nevertheless, we must consider that genetic expression is conditioned by cultural constructions (see later) which notably differ among the various ethnicities. Hence, we cannot avoid continuous changes in the probability of gene expression depending on cultural dynamics. Additionally, the pressing changes generated by the dynamics of scientific-technological developments provide another unavoidable impact on social construction. Hence, from an amplified concept of environmental cues with biological value, they would give another "handle" for differential reproduction.

In our history of civilization, it appears that the social fiction based on collective myths that replaced the world based on objects and concrete events – as stated by Harari (2015) – would not fit nicely into human nature. The constructions of such virtual worlds are the product of collective interactions – corporations, institutions, beliefs, nations, and community organizations. We have not shaken off the ancestral motivational vectors. Additionally, a new world of objects has been generated – instrumental or technological in this case – whose significant development energizes a new cultural push of hard-to-predict consequences in its collective effects, to the point of generating doubts as to whether a posthuman civilization could be budding.

To sum up, based on social repression or "socialization", cultural strata of variable "thickness" have been constructed on top of drives implicit to our animal condition. Nevertheless, it failed in their deactivation, only succeeding in reformulating or repressing them. Consequently, friction between these "geological plates" – of biological and cultural origin –, conditions various aggressive or maladaptive behaviors. In a context of social construction, such evolutive concepts of *territory* and *neighbor to be suppressed*, ought to be replaced by a cultural construction thriving toward cooperative behavior.

The central aim of this book is to offer evidence – to acquire conscience – of our biological nature, evolutive behavioral emergence and our socio-cultural ecology.

It would seem opportune to express uneasiness with the different speed of change between our emotional development as a species, and the vertiginous scientific-technological progress. This *gap* between rational/instrumental knowledge and emotional updating leads to the differential growth of both kinds of intelligence: the emotional and the rational, or cognitive one. Such friction tenses our individual and collective social construction, to the point of generating doubts about the true nature of a hypothetical "posthuman" future (Darlington, 2018).

From having been integrated into natural dynamics, we became imperialists of the macrobiological world, and we developed a collective behavior impregnated with mysticism, megalomania, an unyielding pride of species, and fanaticisms deriving in pugnacious, belligerent fundamentalisms. We have acted in such a way to the point that at least in an active segment of our human community, the inertia of a progressive and expansive technological knowledge has triggered a proposal that pretends to place our roots as marginal to the *Natural Kingdom* (Colombo, 2010). They seem to be fueling the transition through roads leading either to mysticism or a technological blast-off – *technocracy*. Both tend to interfere with our contact with the ground earth; hence, we should have the care to remember what happened to the fate of mythological *Anteo*, to whom his strength abandoned him when losing touch with his mother earth *Gea*, and so was defeated by *Heracles*.

ACKNOWLEDGMENT

To Flavia C. Abdullah, for English language editing.

Chapter 1

THE EVOLUTIONARY IDENTITY

"The genetic divergence time between two species (humans and chimpanzees) varies substantially across the genome, conveying important information about the timing and process of speciation. These unexpected features would be explained if the human and chimpanzee lineages initially diverged, then later exchanged genes before separating permanently" (Patterson et al., 2006).

"Biologically speaking, we are apes that are part of mammalian radiation, primates." "Humans are African great apes" (C.P. van Schaik 2016).

"…This suggests that hominin evolution is neither a simple punctuated process, nor a constant gradual one, but a complex interaction between variable rates of change, environmental dynamics and the competitive interactions of the hominins and their sympatric[2] fellow-travelers in evolution" (Foley et al., 2016).

---

[2] When two species coexist in the same geographic area; sympatric speciation refers to the process through which new species evolve from an ancestral one, while they share the same territory.

1.1. Our Genetic Baggage

According to the *National Geographic,* we share approximately 18 percent of our genome (constitutive pairs of bases of the DNA) with yeast, 24 percent with grapes, 38 percent with worms, 44 percent with bees, 47 percent with fruit flies, 65 percent with chicken, 69 percent with the platypus (duckbill), 73 percent with zebra finch (fish), 84 percent with dogs, 85 percent with cows, 88 percent with mice, and more than 98 percent with chimpanzees. The *Medical Institute Howard Hughes* indicates the following figures of DNA sharing: 31 percent with yeast, 40 percent with worms and 50 percent with the house fly.

It would seem, then, that we do not have too much of a chance to question our pertinence to an almost unbelievable chain of biological events and dynamics, exposed during millennia to unpredictable geo-climatic circumstances. Animals, plants, and fungi share elements of an ancestral genome dated about 1.6 billion years. Each lineage that emerged from the common ancestor retains part of it.

Our construction does not depend only on the DNA pairing of bases but also on regulatory genes. The diverse species of the *Natural Kingdom* evolved through millennia during which additions and mutations and genetic expression depended on natural environmental conditions and the particular ecological niches. According to Kaessman and *Pääbo* (2002) concerning the genomic transference from our ancestors, our genome would consist of about three billion nucleotides that have been passed down to us. Mutations that affect these nucleotides take place in every generation and may accumulate in characteristic ways. Thus, slightly different versions of the ancestral genome are received by subsequent generations. However, embrained in core neural circuits (basal brain mostly) survive our basal responses (fight, flight) and drives (reproductive, territorial, survival, feed).

There would exist a certain consensus (Gagneux and Varki, 2001) that *H. sapiens* shared a common ancestor 5-6 million years ago with the chimpanzee and bonobo, 7-8 million years ago with the gorilla, and 12-13 million years ago with the orangutan. For some authors – based on concepts of genetic anthropology –, chimpanzees would belong to the same genre *Homo* that humans do. According to them, there would be a need for at least a two million year period to support a genetic species divergence; according to this, several fossilized species ought to be considered as *H. sapiens* (see Curnoe and Thorne, 2003; Chaimanee et al., 2012). Given the genetic proximity of the chimpanzee to the human genome, it would require a reclassification in the genre *Homo*, since current taxonomic classification places them as *hominins* (Varki and Nelson, 2007) (see Figure 2).

Osada and Wu (2005) and Patterson et al. (2006) mention that the timing of genetic divergence varies significantly between humans and chimpanzees, which provides information on the timing of speciation. These species (*H. sapiens* and chimpanzees) would have diverged less than 6.3 million years ago, with genomic components that suggest genetic exchange between them following speciation, before final separation. According to Kehrer-Sawatzki and Cooper (2007), DNA nucleotide differences would have ranged by approximately four percent and not one percent. These considerations reveal uncertainties remaining in this evolving field. The split between both species would have taken place some 370,000 years ago, before the emergence of anatomically modern humans.

1.2. *Homo Sapiens* and Great Apes

Human evolution has been summarized (Lahr, 2016)[3] into five periods:

1. Its origin: 240,000-200,000 BCE
2. The first expansions (within Africa): 130,000-100,000 BCE

[3] It seems appropriate to mention the recent advances in detecting *Homo* DNA from sediments, which significantly reduces limitations associated with its detection from fossil remains, and opens up possibilities to complete the evolutive map of the genre *Homo*. Researchers have analyzed sediments from seven archeological sites covering an approximate period of 550,000-14,000 years (Slon et al., 2017).

3. Initial dispersion (out of Africa): 70,000-50,000 BCE
4. Local configuration of diversity: 45,000-25,000 BCE
5. Significant extinction of hunter-gatherer groups and expansion of farmers and subsequent cities (the Holocene filter): 15,000 BCE

To this sequence, it should be added the gradual balkanization process of humanity concerning access to knowledge, quality of life, technology, and education. This has taken place progressively through the various forms of social inequity and access to knowledge, and accelerated during the few last hundred years; a process that will mark significant differences in human community development and tense relationships among its components.

The impact of human activity on the environment promoted the addition of the *Anthropocene*[4] era, a signal of the level that human activities have attained – a history that emerges from the haze of ancestral evolutive hazards as members of a taxonomic group called *hominins*, which we share with great apes, within the Primate order.

The genetic differences among hominins do not always emerge clearly (Patterson et al., 2006). According to Foley et al. (2016), *hominin* evolution was not an event of a punctate process nor a gradual one, but the product of a complex interaction with variable rates of change depending on environmental dynamics and competitive interactions with other contemporary *hominins*.

Besides providing a taxonomic classification, the issue remains to explore the vast number of hominids and hominin lineages, and reconstruct our inherited fundamental condition that is camouflaged behind our sophisticated and complex diverse cultures. In this sense, Preston (2013) provides a statement that involves a significant impact on our self-conscience of our cultural construction:

[4] Anthropocene: geologic era proposed to include the impact of human activity.

> "…even complex human behaviors reflect ancient mammalian neural systems that evolved to solve key problems in adaptive ways, with far-reaching consequences for even our most venerated human traits."

What comparative evidence contributes to genetic studies regarding brain development and organization? Dorus et al. (2004) identified a group of 24 genes[5] involved in brain development that differed in humans and rodents. A particularly interesting one is the *ASPM* (Abnormal Spindle-like Microcephaly-associated protein) located in chromosome 1. It codifies for a protein involved in cell division, particularly in the central nervous system, which has an impact on cerebral cortex growth (Rakic, 1995a, b; Rash et al., 2016). Control of this cell process results in significant differences in the final growth of the cerebral cortex (Chenn and Walsh, 2002). According to Evans et al. (2004), a positive selection over the *ASPM* gene would have started around 18 million years ago and remained until the emergence of the modern human. From the standpoint of social and biological interactions, it is interesting that mammalian species with the least cortical proliferative index tend to live in reduced social groups, compared to those species showing a broader index of cortical folding (Lewitus et al., 2014; Weaver, 2014), i.e., composed of a more substantial number of cortical cells.

The evolutive impact on cerebral cortex growth would not depend only on population size, but also on the complexity of its social structure (Dunbar and Schultz, 2007a). According to these authors, the social brain hypothesis (with emphasis on the relative size of the neocortex concerning the total brain) represents an ecological hypothesis, according to which one or more ecological problems (survival, access to nutrients, raising of offspring) are better solved in a social context than individually.

Another gene probably associated with human evolution is *FOXP2*, presumably involved with language development. Comparative studies including great apes and Macaque genes show changes in amino acid

---

[5] *Genes*: an informational unit in a locus of the DNA that codifies for a given functional product; a molecular (nucleotide sequence) unit of genetic heritage.

codification and nucleotide polymorphisms, suggestive that it has been a nodal event during recent human evolution (Enard et al., 2002). Within this context it is interesting the remark made by Gagneux and Varki (2001):

> "...it is much more probable that our evolution occurred in multiple genetic steps, many of which must have left detectable footprints in our genomes."

The extended construction period of the cerebral cortex (cell production and migration) and of its organization regarding intracortical and subcortical connectivity makes it particularly labile to environmental agents and conditions. This issue – its functional consequences on cognitive and emotional development – has been dealt with in other instances related to socioeconomic conditions (poverty, indigence, marginality). Such a period of susceptibility to external factors makes it permeable to positive and negative events on the brain and mind construction (see in Colombo, 2007; Lipina and Colombo, 2009; Lipina et al., 2012).

1.3. Sapiens Evolution: A Multifactorial Issue—Gut Microbiome

> "If we consider ourselves to be a composite of microbial and human species, our genetic landscape a summation of the genes embedded in our human genome and microbiome, and our metabolic features a coalescence of human and microbial traits, the self-portrait that emerges is one of a 'human supraorganism'. Thus, understanding the range of human genetic and physiologic diversity means that we must characterize our microbiome and the factors that influence the distribution and evolution of our microbial partners" (Turnbaugh et al., 2007).

Several hypotheses and theories have been proposed to understand the brain and mind development of *H. sapiens*. Among them are social complexity as an ecological factor (Dunbar, 2009; Dunbar and Shultz, 2007b); a shift in feeding behavior with the subsequent reduction of the digestive system in order to compensate energy demand due to brain growth (Aiello and Wheeler, 1995); genetic modifications (see before); neurotransmitter involvement, such as dopamine (Previc, 1999, 2009); and culture (see ahead). These proposals do not seem to contradict, but rather complement each other, and acting either at similar or different (sequential) times.

It would be an unjustifiable omission not to mention another factor that has been usually omitted from the multifactorial equation behind *sapiens* or *Homo* development. The consideration of such a factor – the *gut microbiome* – will be expanded here and linked to existing differences among primate species, as well as with brain development and behavioral profiles. Its role in our physiologic integration has led to consider our organism as multi-genomic, that is, in addition to our endogenous DNA, we possess a universe of bacteria that coexist in our organism.

According to Díaz Heijtz et al. (2011), gastrointestinal colonization by microbiota would have become integrated with the programming of brain development, and involved in motor control and anxiety-like behaviors. As Buchanan and Bohórquez (2018) have lucidly mentioned, the relationship of the gut with developmental issues precedes postnatal life. In this respect, amniotic fluid components – whose composition depends on the mother's habits – circulate after being swallowed by the fetus. It is the first medium that the gut receives; how this sets up future functional interactions remains to be further studied. Interestingly, the human amniotic fluid has been shown to foster the growth of rat neural brain cells in culture, an effect disrupted by antibodies to nerve and epidermal growth factors (Colombo et al., 1993). Hypothetically, the increased permeability of gut walls at early stages of fetal development (Robertson et al., 1982; Weaver et al., 1984) would allow amniotic components to enter the humoral media and act as a trophic brain component. In this regard, it is worth mentioning the experimental observation by Colombo et al. (1993):

> "It is concluded that amniotic fluid constitutes a trophic medium for astroglia and neurons".

Based on previously mentioned data, it now could be added that this neural trophic activity may operate during fetal development.

Following a comparative study, Ley et al. (2008) concluded that diet and phylogeny regulate gut microbiome. Within…this context, the authors state that its diversity increases from carnivores to omnivores to herbivores and, in humans, it will be typical of omnivore primates. From a metabolic point of view, such microbiome would codify for functions that our genome would not have developed entirely on its own, including the capacity of extracting nutrients and energy from our diet (Ley et al., 2008).

According to Rieder et al. (2017), the gastrointestinal tract houses more than 100 trillion microorganisms, which can weight up to 2 kilograms and contain at least one thousand different species of bacteria. This microbiota represents the ecological community of commensal, symbiotic and pathogenic microorganisms.

If we put into perspective these considerations with respect to the possible metabolic impact of the gut microbiome (Tremaroli and Bäckhed, 2012), and on the debated role of metabolism and energy requirement on brain development and body size (Kleiber, 1947; Armstrong, 1983, 1990; Leonard et al., 2003; Glazier, 2005, 2010; Navarrete et al., 2011; Cryan and Dinan, 2012; Fonseca-Azevedo and Herculano-Houzel, 2012), it would appear that gut microbiome represents a potentially significant factor among those involved in order to understand species evolution, including ours.

Besides its potential impact on the species' evolution, postnatal food and gut microbiota have been implicated in brain function and dysfunction. The effect of the microbiome on the nervous system has been studied concerning development and behavior. It is important to remember that it establishes a close symbiotic relationship with the host since birth, although it goes through successive changes due to shifts in the feeding habits as we grow up.

According to Borre et al. (2014), perturbations in such feeding sequences and conditions could affect development and adult behavior. From an experimental point of view in laboratory animals, it has been possible to confirm that the microbiome could modulate behaviors related to social interaction, exploratory activity, and anxiety. It would also affect the expression of neurotransmitters (GABA) and neurotrophic factors (*brain-derived neurotrophic factor*-BDNF) (Díaz Heijtz et al., 2011; Bercik et al., 2011; Bravo et al., 2011; Borre et al., 2014; Buffington et al., 2016). As an example, in humans, the probiotic *Bifidobacterium longum* would have an impact on anxiety disorders associated with gut dysbacteriosis, probably affecting neural systems related to the limbic system (Pinto-Sánchez et al., 2017).

The mentioned studies show an impact of the gut microbiome on the structure, physiology, and development of the nervous system and behavior. The probable mechanisms that link the gut microbiome with the nervous system have not been established, although the vagus nerve, the immune system, and the humoral route, could be involved. In this context, it is of interest to speculate on the possible role of feeding habits on the evolution of *Homo*. Additionally, these considerations should be included in the construction of public policies regarding child nutrition.

Chapter 2

EVOLUTIVE IMPACT OF NEUROBIOLOGICAL AND CULTURAL DEVELOPMENT: EPIGENESIS

"Much additional data concerning the behavior, cognition, physiology and pathology of chimpanzees and other NHHs (nonhuman hominids) has since accumulated, showing how remarkably similar we are, and yet how different". ... "Ultimately, it is necessary to understand uniquely human aspects of phenotype in the context of genotypic differences from the non-humans (NHHs). Many lists of such differences have been published, and it is becoming increasingly clear that several of these differences are relative rather than absolute" (Varki et al., 2008).

2.1. PRELIMINARY CONSIDERATIONS

Africa, year 200,000 BCE. A group of anatomically modern *hominins* carrying new fighting tools and strategies was competing for food with other groups. For hundreds of years, their descendants migrated and spread throughout the planet; in some places where other *Homo* species were already inhabitants (*H. erectus, H. neanderthaliensis, H. denisovan*) and where they possibly exchanged genetic material. Behind them, spread in time and space, remained fossilized remains of a multitude of species either predecessors or phylogenetically related to the diversity of the genre *Homo*. Their physical survival possibilities were exhausted but intermingled with the genome and biological organization of the new, prevailing species. Scholars are now finding traces of what in all probability represents a formidable, dispersed cemetery of previous survival trials and a sort of "genetic haze" in our evolutive trail.

As stated in a *Nature* editorial (2016):

> "Were we able to mark the remains of all our ancestors from that point, the world would be one enormous cemetery." "… provide yet more evidence that the human lineage is more diverse than was ever imagined."

Descendants of such a genetic pool – species comrades but doctors in anthropology, molecular biology, genetics, paleontology, and other related disciplines – analyze today those scarce and hidden fossilized remains of our large, ancestral family. Moreover, where their remains invaded the haze of time, there persist traces of their DNA footprints in the soil (see footnote three referenced in section 1.2).

Social cohesion, inventiveness, efficient strategies, accumulation of knowledge – memory as the engine for change –, and communication prevailed. With what evolutionary backpack did they evolve initially? What deep traces were imprinted in their brains and projected on to their minds regarding the legacies of the long trail of experiences built on successive evolutive stages? Did such traces leave their marks associated with their condition of prey and predator, facing cataclysms, terrifying climactic phenomena, sickness, fears, failures? What ancient biological inertias motorized their development, and today underlie our behaviors and modern constructions, whether symbolic or virtual – beliefs, dogmas, norms, values?

The biological nature of our construction drags millenniums of trials with variable rates of success: of successful species during extended periods and others with comparatively short or unknown survival success. All of them provided traces of a variable and multiple evolutive chains.

As in all fields of scientific knowledge, there coexist different hypotheses, theories, and proposals, in this case, related to the possible "evolutive path" of the genre *Homo* and *sapiens* species. It is possible that our direct ancestor *Homo* would have been the *Homo erectus*, or variations that emerged from it (*Homo antecessor*, *Homo heidelbergensis*). This long-lived species would have made successive migrations within Africa and toward the Euro Asiatic continent, leaving enclaves that were later invaded in different regions by other *Homo* migrant species, including the *anatomically modern sapiens* and Neanderthals. With time, such migrations and interbreeding, coupled with ecological adaptations, contributed to the generation of different phenotypes and ethnicities. In an editorial commentary of the journal *Nature* (2016), it was reported the vast variety of hominids that preceded our species, some of them occasionally cohabiting with the modern *sapiens*, thus, forming a large collective to a great extent relegated from public understanding of the evolutive circumstances of our species evolution.

Foley et al. (2016) and Lahr (2016) summarize the possible stages in the evolutive process of the *hominids*, with particular emphasis in the transitions resulting from reductions in the demographic density of various lineages. Changes in primeval demographic distribution could have propitiated the emergence or differential expansion of other lineages, hence generating transient stages in the evolutive display.

As we may see, theories regarding the hypothetical evolutive chain leading to our species may vary and recognize different routes and possible genetic exchanges (Kuhlwilm et al., 2016). Where there is no doubt is that our origin emerges following a succession of hominid species, each one of them labeled by a structural component and functional and survival strategies (Foley et al., 2016; Lahr, 2016; Kuhlwilm et al., 2016). Our closest ancestors evolved from such a variety of lineages or an evolutionary framework, in which epigenetic circumstances completed the speciation of the modern human.

Interaction with the physical and cultural environment continues modeling our ethnic variations, yet, our primary organization is bound to ancestral demands that imprinted a given set of basic behaviors, as mentioned before (territorialism, sexual drive, individual survival, secure feeding sources). Their expression, affected by changed environmental (physical and social) conditions, pose the probability of frictions between the neurobiological and cultural *tectonic plates* (Colombo, 2010, 2013, 2016). It is true that the plasticity of our brain and mind construction (depending on cultural issues) provide room for adaptative responses. However, they do not cancel the framework of primary drives imprinted in the heart of our animal construction, although they may affect their expression. Psychological mechanisms disguise, or temper, some of the behavioral consequences of such frictions or uncoupling.

In all gregarious species, the reiteration of social organization patterns – of relationships between individuals, the group structure to which they belong, and the environment – end up reconfiguring the neurobiological substrate supporting individual behavior. It is apparent that, at some point in time, in the genre *Homo* such a process induced a progressive change in the increasingly complex social demands placed on the neurobiological organization, resulting in the *cerebral incarnation* of culture. This implies an additional demand on brain/mind processing that exceeds the cerebral incarnation of the basic rules/codes present in every non-human gregarious animal community with a hierarchical organization. In *Homo*, this process must have been slow due to its evolving complexity, progressively based on values, beliefs, language, symbolic constructs, productive chain, neoteny of the newborn, coupled with other survival conditions (feeding habits, social organization, sedentarism, reduction of the stress evoked by risks due to survival as a prey); a process that allowed the brain to undergo microstructural changes affecting neural circuits associated with its social organization. At the same time, dynamic biosocial interactions conditioned the genetic expression.

According to some authors, such a cultural development would have had a long, slow, progressive format. However, according to others it would have been an accelerated one spanning 40,000 years – the so-called cultural revolution – (see Renfrew, 2008), probably associated with the emergence of language. Following McBrearty and Brooks (2000), the time lag generated between the appearance of the *H. sapiens sensu strictu* dated more than 100,000 years, and the claimed emergence of modern human behaviors known as the *human revolution* ca. 40,000-50,000 years ago, creates the notion that the earliest modern African individuals were behaviorally primitive.

This apparent uncoupling between the genetic, physical and cultural configuration of the *anatomically modern H. sapiens* is represented by what has been called the *sapiens paradox* (Renfrew, 2006, 2008). What neurocognitive modifications must have taken place in order to develop a symbolic expression and what circumstances have contributed to it? It is interesting the temporal coincidence of parietal graphics and artefactual expressions, and evidence of ritual behaviors. For some authors, this suggests an interactive process with a brain physiologically apt to associate distributed neural circuits. In such a process, phenomena of *exaptation* – when neural circuits or structures associated with specific functions are later recruited for other functional means – and of *emergency* – when new connectivity is involved, whether implementing structural or reinforced synaptic pathways –, take place. On this issue, both gradual and revolutionary theories collide.

According to Ian Tattersall (2004, 2017):

> "The extraordinary cultural achievements of the Upper Paleolithic Europeans show that by 40 kyr ago the exploitation of the human symbolic potential was well advanced; but wherever and whenever this highly generalized capacity may have been initially discovered, substantially after the birth of *Homo sapiens* as an anatomical entity, the discovery of how to use it in new ways was a gradual process and one that is still proceeding" (2004).
>
> "Nothing indicates convincingly that any hominid behaved symbolically prior to *Homo sapiens*, which originated 200,000 years ago but started behaving symbolically only 100 k yr. later. Most probably the necessary neural underpinnings arose exaptively in the extensive developmental reorganization that gave rise to anatomically distinctive *Homo sapiens*, and were recruited subsequently via a necessarily behavioral stimulus" (2017).

Within these sets of proposals, one would need to include the concept of synaptic Darwinism (Adams, 1998) according to which neural connections are canceled or reinforced, or generate conditional new ones. That is, this proposal combines those concepts of Darwin's theory applied to species evolution, and of Hebb (1949) applied to synaptogenesis, synaptolysis and synaptic reinforcement at the microstructural level.

It seems opportune to remark that the expression and dynamics at the molecular level of neurotransmitters, neuromodulators, and receptors, together with the bioelectric process, are the basis of synaptic connectivity and escape entirely from any chance of being identified by paleoanthropological, paleogenetic and archaeological procedures. Due to their very nature, they generate an "anthropological gap" built by their short survival span and, hence, lack fossilized remains. On these grounds, it is opportune the call by Colin Renfrew (2008) to undertake interdisciplinary studies between paleoanthropology and neuroscience.

At the comparative level – another avenue to study these phenomena – it is worth quoting the comparative work performed by Karl Zilles and associates (Zilles, 2005; Zilles y Palmero-Gallagher, 2017) on the cortical distribution of neuroreceptors and neurotransmitters, appealing to extant anthropoid species. In this regard, it is interesting to remark the human differences observed with the Macaque (Zilles et al., 2005, 2007) concerning density and distribution of receptors to a series of potential neurotransmitters in the prefrontal (orbital) and posterior parietal cortices. The latter perhaps associated with modifications involved with the use of instruments, and attention related to egocentric and allocentric spaces (Zilles, 2005). Observations on both cortical regions stress the differences in the associative nature between the human and Macaque brains.

2.2. Evolution, Epigenesis, and Neurobiological Events

Let us go back to the emergence from Africa of the *anatomically modern human* in successive migration waves some 200,000-300,000 years ago. They progressively replaced or merged with archaic forms of *Homo* residents in Europe and Asia, a product of previous migrations from Africa of the *Homo erectus* some two million years ago (Kaessmann and Pääbo, 2002), with whom they would have had an opportunity to exchange their genetic pools. Although at present the taxonomy of the great apes is still a matter of discussion, several million years earlier, speciation processes had differentiated *Homo* from them.

According to Varki and Altheide (2016):

> "Humans and chimpanzees shared a common ancestor 5 – 7 million years ago (Mya). The difference between the two genomes is not appr. 1%, but appr. 4% - comprising appr. 35 million single nucleotide differences and appr. 90 Mb of insertions and deletions."

Given the relative genetic proximity, it seems of interest to point out some differences observed between humans and chimpanzees concerning some events of nervous system development, as compiled by Varki and Altheide (2005) (Table 1). In this same context, according to Varki et al. (2008), the evolutionary relatedness of humans and great apes[6] was predicted by Huxley and Darwin, supported later by molecular studies.

Several authors (Curnoe and Thorne, 2003; Chaimanee et al., 2012), support the concept that it would be necessary periods of two million years to sustain a species' genetic divergence, according to which several fossil *hominid* species should be considered as *Homo*, hence, species phylogenetically related with ours. Such biological proximity – relatedness – with a series of hominid ancestors seems unquestionable, according to genetic and paleontological research. Let us remember that our genome represents the accumulated and modified genetic pool from ancestors to our species and posterior genetic exchanges – confirmed or potential – with different species within the genre *Homo*, settled in different regions of

---

[6] *Hominid*: a clade that includes humans and great apes (chimpanzee, bonobo, gorilla, orangutan). *Clade*: group of organisms with the same ancestor and its direct descendants; they represent a "branch" of the "tree of life".

the Euro-Asiatic continent (*H. neanderthaliensis, H. denisovan, H. erectus*).

Table 1. Differences observed between humans and chimpanzees (after Varki and Altheide, 2005)

Neuroanatomy
Relative Brain Size
Direct Cortical Projections
Relative Volume of Frontal Cortex
Relative Volume of Corpus Callosum
Relative Volume of Cerebellum
% of Brain Growth Complete at Birth
Rate of Postnatal Brain Growth
Neurobiology
Population Distribution of Handedness
Postnatal Dendritic Growth
Postnatal Synapse Formation
Cortical Synapse Density
Cortical Neuron Density
Dendrites Per Neuron
Synapses Per Neuron
Adult Neurogenesis
Fingertip Sensory Nerve Endings

2.3. Culture[7] as an Evolutive Modifier. Construction of Identity

The role of culture as the environmental configuration factor most proximal to individuals has been stressed by several authors (Boyd and

[7] *Culture* (material, symbolic, social) is a differentiated collective practice, transmissible to other group members systemic and adaptative, a practice that includes some sets of shared significance. See in Kitayama and Uskul (2011).

Richerson, 2009; Kitayama and Uskul, 2011). The bio-cultural interaction that ensues has a double impact. The culturally evolved environments affect which genes are favored by natural selection and, in turn, as stated by Kitayama and Uskul (2011), genetic characteristics affect the expression of certain forms of culture instead of others.

In this bio-social interaction, the brain/mind acts as depositary of both acting vectors; that is, the genetic inheritance expressed in brain organization and its biological limits, and as the generator and depositary (receptor) of cultural events that modulate such expression. These mutual conditionants of individual and collective phenotypes have been in action through centuries, with different structural complexities of the gregarious group interactions. The repetition of individual and collective behavioral patterns tended to crystalize basic behavioral configurations, with an added brain/mind plasticity in the case of humans. The latter probably due to the emergence of a language that fostered virtual interoceptive and exteroceptive constructions. This virtuality expanded the *degrees of freedom* by which an individual relates to the physical and emotional environment.

In previous instances, we have mentioned some of the neurobiological mechanisms related with the concept of neuroplasticity and learning, such as synaptic formation and elimination (synaptogenesis and synaptolysis), neurogenesis, reinforcement of synaptic connections, synaptic potentiation. These processes represent the biological basis for the encephalization of cultural events when accessed in a reiterative fashion – *brain incarnation of culture* –, and are incorporated as microstructural and functional modifications in brain organization (Kitayama and Uskul, 2011; see at the experimental level in Teles et al., 2016; in humans using functional magnetic resonance, see Schwartz et al., 2002).

Previous paragraphs would serve as a general introduction to the possible impact that cultural construction would have had on the biological evolution of humans. As mentioned, the hypothesis linked to the particular evolutive increase of brain volume and the complex organization of brain and mind in modern *sapiens* species has taken base in a series of biological and social considerations. Among them are: a change of basically frugivorous feeding habits to an omnivorous diet, which would have compensated for the energy expenditure of the required high metabolic rate of the brain at the expense of the digestive system (*expensive tissue* hypothesis, Aiello and Wheeler, 1995); radial (columnar) brain expansion of the cerebral cortex, allowing for its folding, instead of a vertical expansion which would have further complicated in terms of brain function – length of cell processes – and fetal delivery viability due to the cranial volume it would have implied (Rakic, 1995); savings in terms of connectivity due to relative white matter reduction, based on brain modular organization (Ringo, 1991); postnatal extension of the time for brain development (principally cortical) or *neoteny* (Couqueugniot et al., 2004); demands imposed by social complexity on brain computational power (the *social brain* hypothesis) (Dunbar, 2009), further expanded to include ecological concepts – tool making, foraging – and transform it into an *ecological brain* hypothesis (Reader and Laland, 2002; Reader et al., 2011); cultural impact on mind construction and brain adaptation leading to a *cultural intelligence* hypothesis[8] (van Schaik et al., 2012) or of the *socio-cultural brain* (Whiten and van de Waal, 2016); and involvement of subcortical white matter distributed circuits in brain organization (Colombo, 2018).

Earlier commentaries on the bio-social interaction (interactive conditioning from the socio-cultural and genetic domains) lend support to an integrated concept of the human brain and mind evolution. As we see, these concepts do not cancel each other, but instead form the basis of the compound, multifactorial equation involved in *sapiens* evolution, to which we should add language development and with it the symbolic, virtual, construction and a semantic (logical and lexical) potentiation of thought construction.

---

[8] It is interesting to note that van Schaik et al. (2012) point to a couple of shortcomings on an exclusively "social" hypothesis of brain development, based on the observation that several primate species and lineages do not express the expected brain size for their social complexity, and that species with manifest socio-cognitive special abilities also express general cognition.

Our species was not born in a mother-of-pearl cradle and protected by magical agents. Hence, let us recognize that our species' origin finds its roots in ancestral habits, behaviors, and survival drives, that during millennia were crystallized in basic neurobehavioral circuits, be it as predators or potential preys.

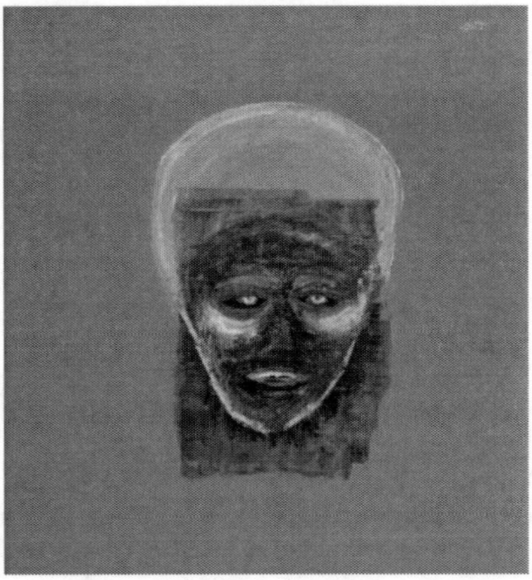

Figure 3. Artist's view (J.A.C.) of the human "bipolar" condition. (Modified into grey tones from the original by J.A.C., 2013.)

Placed on the thread of time, modern cultural components – norms, priorities, values – appear as newly born and of recent construction. This sort of interactive dystopia between ancestral drives and modern social demands results in an arch of possible behaviors, whether adapted, confrontative, disruptive, sublimated into creative drives, or pathologic. Would there be room for reprogramming both sides of the equation and reach a new productive equilibrium? It remains as a debt of public policies leaders and scholars to explore it and attempt it.

Within this context, aggressiveness and Machiavellian behavior between interpersonal relationships, or among competing communities and States acquire "permissible" forms or often reach a dangerous condition where the *sapiens* culture appears threatened, as shown through modern history. The bottom line of such dynamic remains adhered to certain basic components of our ancestral behavioral drives (power prevalence, territory, food security, parochialism, genetic heritage), although camouflaged under various cultural costumes, some of which tend to exaggerate basic drives (accumulation of material and power).

The cultural slope of technological development as a differential developmental factor among communities has continued its growth and permeated most cultural niches, although with different cultural goals and results; thus, invading our daily habits to become another *brain incarnation* with different applications depending on specific segments of the human community. Although technological development has potentiated various aspects of the cultural framework since the lithic period, present time dynamics of technological advances introduced a significant variable of adjustment in specific social strata, approaching a state of *technocracy*. That is, technological and instrumental applications govern our priorities and behavioral profiles. This would have taken place since the first notion of "instrument" entered our socio-ecological niche. Its molding-behavior power has entered a steep slope that, at present, shakes our very basic structure of values and cultural scaffolding. Humane relationships are menaced by technological advancements – threatening to become *technocracy* –, or at any rate, it is reformulating them.

Our emotional intelligence has lagged with respect to the options opened by scientific-technological developments. The latter has set up the pace of the modern human culture in post-industrial communities where an electronic interlocutor and technocratic objectives have replaced human presence. Regarding interpersonal communication, technology has provided means for laconic and lexically and semantically compressed forms of communication that tend to replace access to gestures and voice modulation, placing the notion of "*state of mind*" – or *Theory of Mind*, a primary component in human communication – in the context of *a neurocognitive process at risk of extinction*.

Should we place into perspective previous comments on bio-social interaction (see Cole, 2009), it appears as predictable the progressive *encephalization* of technocratic culture, and the gradual birth of a *posthuman civilization* (Darlinton, 2018). This would primarily anchor in certain community strata, as a fertilized ovum which found its bed in ripe uterine tissue. It is to be expected that this preferential adherence – *embrainment* – of *technological culture* to culturally primed social groups/strata will have its impact as an epigenetic[9] factor. Thus, it would differentially affect genetic expression/inhibition.

On comparative grounds, given the abysmal developmental differences among world communities – which questions the real meaning for the publicized concept of *globalization* –, it is conceivable that privileged or technologically advanced societies – or groups among them – would change their dynamics. Consequently, foster the growth of a culturally/technologically balkanized world. That is, a grand *divide* between scientific-technologically advanced societies – and eventually communities within them – and those that are not.

Moreover, it will drive other events and issues in social construction. The flaw of the concept of globalization in an inequitable and balkanized world community regarding access to knowledge and critical access to information, will become more apparent, deepening the present process of *segmentation* of the human community.

On ecological grounds, the uninhibited voracity for energy sources and the search for new essential elements for technological developments, suggest that there will be a continued break of the limits for ecological prudence. In such eagerness, the advantages of mineral exploitation in meteors and planets has already been voiced (Elvis, 2012); after all (*sic*) "*greed is a powerful motivator to get things done*" (Elvis, 2012). A real shame – and worrisome future – that such projects would be based on *greed* rather than on human values and collective evolution.

[9] Research on social genomics has established that our extrapersonal world exercises significant biological effects on the molecular composition of our bodies (Cole 2009; Slavich and Cole, 2013). In such sense, it has been shown that the DNA not always codifies for protein synthesis, and that it is possible to activate or inhibit the expression of genes due to social and environmental conditions (such as social isolation, inadequate maternal care), by affecting DNA methylation and changes in his tones that may affect activity and message transcription, thus generating potential long duration or late effects (Meaney and Ferguson-Smith, 2010; Robinson et al.,2005; Robinson et al., 2008; Tsankova et al., 2007; Weaver et al., 2004; Weaver et al., 2006; McGowan and Szyf, 2010a,b; Boyce et al., 2012; Fernald and Maruska, 2012).

Nothing seems to reassure that our world social organization will gain social equity and solidarity. Concerning profit and human development, the *overspill of the full glass* seems to operate in a desperately slow way and subordinated to other corporate and political priorities. Also, this poses another question mark. According to the news in public media, private corporations are developing means to explore the extra-terrestrial space. Hence, would its capacity – financial power – for the exploitation of resources, be on hands that search for profit (*greed*) rather than for human interests? There seems to be a profound need to change cultural gears on the motivation for the political use and promotion of technological progress.

Segmentation in the access to knowledge, to informed behavior, and creativity seems to be adding new developmental profiles to the balkanization of the human community. Attempting to define the phenotype of the technological *Homo* exceeds the nature of this book, specifically, what could be its consequences on the individual and collective emotional construction. We can add that in present conditions, access to valued information is restricted/conditioned due to economic, socio-cultural, and political reasons, all of which generate profound differentiated strata of access to information, and educated, critical, access to it. For actual communication implies access to data and an educated receptor to make sense out of it.

Within that informational void and technological-generated anomia, we can sense a proliferation of mysticism, esoteric and ritualistic thoughts and practices that would attempt to overcome handicapped cultural conditions, ecological and geo-climatic disasters, personal misfortune, social marginality, or belligerence fed by religious fundamentalisms.

2.4. Preliminary Comparative Issues: Cooperation/Competence

Cooperation is the basis of the organization of multicellular organisms. Cooperation based in reciprocal altruism would have evolved only in some species (Axelrod and Hamilton, 1981), yet it constitutes the fundamental -founding- principle of human society.

Cooperative interaction among community members is a significant factor involved in brain and mind evolution, and individual and social development. According to Burkart and van Schaik (2016), cooperative child upbringing promotes proactive, prosocial behavior. Such cooperative breeding – linked to shared intentionality – is also expressed in other species (Bukart et al., 2009). Regarding the potential environmental impact on biosocial interaction, it is worth insisting that social and environmental conditions can affect gene expression and epigenetic processes – DNA methylation, histones (proteins of nuclear chromatin) modifications (see also Slavich and Cole, 2013).

According to Boyd and Richerson (2009), during the *Pliocene* era (3-5 million years ago), societies of our ancestors were mainly similar to those of other primates (also in Foley and Gamble, 2009). According to the socio-ecological view of these authors, the most notable changes in social evolution would have been modulated by changes in the ecological relationships among humans, their culture, and natural resources. Such changes would have led to a generation of new behaviors and a greater control of these over the construction of communication nets, social superstructures, and the distribution of labor.

On a basic level, structural parameters that characterize human societies – with significant cultural variations – tend to be similar to those of other vertebrate species: the tendency to form social groups, prolonged parental relationships, different tolerance to within and foreign groups, hierarchical structured social forms. According to Ghiglieri (1987), the fundamental ancestral organization of human society would have been based on several males and females, a behavior that would have evolved from the *Pliocene* (between 2.6 and 5.3 million years ago) *ancestral Pan*, common to the evolutive branch of chimpanzees, bonobos, and *Homo*. Social evolution would have adopted new community structures, according to the mentioned author and Boyd and Richerson (2009), during the *Pleistocene* (10,000 to 2.5 million years ago). A distinctive trait in this process would have been the capacity to learn from other community members and to transfer such learning. This expresses a psychological characteristic that would have differentiated *H. sapiens* from other phylogenetically close species. According to Boyd and Richerson (2009), this would have evolved as a cultural adaptation possibly associated with sudden climate changes during the *Middle Pleistocene* (130,000-780,000 years ago, approximately). Such a process would have resulted in cooperating/competing groups generating a new dynamic growth among them, improving their survival and the expression of genes involved with prosocial behavior. Pradhan et al. (2012) include technological or instrumental evolution, which is considered a threshold in cultural evolution, mostly starting from the *Acheulian* lithic period (0.1 to 1.76 million years ago) of the *Lower Paleolithic* (approximately 300,000 to three million years ago).

Intraspecific cooperative trends among *H. sapiens* would have established an evolutive advantage over other *Homo* species. However, are cooperative and competitive behaviors a specific characteristic of humans? Some comparative studies help to build up a better understanding of this issue.

Moll and Tomasello (2007) concluded that infants from human and non-human hominids share the ability to understand others as actors with objectives. While non-human primates display their abilities with the highest efficiency under competitive conditions, human infants seem to display special abilities in cooperative tasks with shared intentionality. Authors shared the *Vygotskian Intelligence* hypothesis, according to which cooperation with shared intentionality represents in humans a non-primitive derived characteristic, which generated the emergence of a cumulative cultural evolution, with its symbols and artifacts. According to Hare and Tomasello (2014), chimpanzees are more skilled at tasks that require competitiveness, as opposed to cooperation. When chimpanzees interact with tolerant partners may behave as cooperative. These authors (Hare and Tomasello, 2004) suggest that human cooperation would have emerged following a process of eradication of the less cooperative members.

According to Moll and Tomasello (2007)[10], based on concepts of N. Humphrey and L. Vygotsky, the primary engine for human cognitive evolution has been competitiveness and cooperation, respectively. Unlike non-human primates, from early ages, humans start to incorporate forms of cognitive representation, potentiated by the comparatively prolonged postnatal neurobiological immaturity period (*neoteny*). Although the complex constellation of characteristics of the human mind is not shared in the animal kingdom, nevertheless homologies and analogies of critical aspects of the human mind can be observed in several animal species (MacLean, 2016). Within this context are particularly interesting the comparative observations made on patterns of cooperative prosocial behavior between chimpanzees and bonobos – both descendants from the *ancestral Pan*. While the former shows a social organization based on male hierarchies, greater inter-group aggressiveness and limited cooperation with less interindividual tolerance, the role of females in bonobo tribes is predominant and show less aggressive displays. These differences have been attributed to habitat characteristics and of their proximal competitors and allow bonobos a better performance in cooperative tasks where the prizes can be monopolized.

From a neurocognitive view, it is interesting that Rilling et al. (2012) have observed connectivity differences among some brain areas (amygdala, insula), supposedly associated with differences in the basic behavioral patterns of the bonobo and chimpanzee.

Humans share with closest primate extant species a series of anatomical and functional characteristics (Sallet et al., 2011; Dunbar, 2003; Flinn et al., 2005; Roth and Dicke, 2005). The relative increase of cortical grey substance – and its characteristic neuronal density and connectivity – has been associated with the process of socialization and growing social complexity in primates, in particular within the genre *Homo*.

---

[10] According to these authors, although both coincide in the social role as the engine of cognitive evolution, Nicholas Humphrey places the accent on competitiveness, while Lev Vygotsky places it on cooperation. It is apparent that from an ethical point of view, competence may promote behavioral deviations with negative social impact, which do not appear as evident under cooperative conditions

Following the previous reports, characterization of those species as being either cooperative or competitive would depend on the conditions studied. Nevertheless, the predominant characteristic in each of them differs, and that is translated to their social structure and dynamics. According to De Waal (2005), we humans would be bipolar:

> "What is most critical for survival: competition or cooperation? Are we more like chimpanzees or bonobos?... Such questions are a waste of time for the bipolar characters that we are."

If the expression of this supposedly bipolar human nature depends on cultural conditions, the occasional predominance of one or the other would depend on public policies that contribute to the construction of social relationships. Besides, if it is considered that behaviors are induced or reinforced by its reiteration, such public policy decisions acquire a collective pedagogic significance with an impact on the future. That is, the construction of a society where such bipolarity could be reduced and express itself as either cooperative or competitive.

2.5. Epigenesis and Socio-Cultural Factors

The nature of factors that could act as epigenetic[11] agents – capable of modifying DNA expression without affecting its structure – include chemical substances, nutritional factors, stress, temperature and those cultural conditions sustained in time that result in the promotion or inhibition of gene expression, and contribute to defining phenotypes and ethnicities. Let us remember that besides heritage associated with the gametes, there is the DNA expression at the cellular level in several brain

---

[11] Posttranscriptional mechanisms (involving ribonucleic acid, RNA) in genetic expression. A great deal of research on epigenetics has been concentrated on marks associated with DNA, but more than 100 marks were identified in the RNA (epitranscriptomics) (Willyard, 2017).

structures, i.e., that affect the expression of the cell heritage. These changes could affect an individual's behavior but will not be heritable.

It must be added that persistent changes in the physical, cognitive and social environment produced by adaptative cultural inertias can induce phenotypic modifications – cultural heritage – probably expressing preferential genetic expression. In this sense, it is interesting to mention observations made by Jablonka and Lamb (2007), Cole (2009), Lubin et al. (2011), Graff and Tsai (2013), Rudenko and Tsai (2014), on such epigenetic modifications with impact on cognitive functions and the synaptic substrate of the learning and memory processes.

McGowan and Szyf (2010a, b) have analyzed possible dimensions of the concept of epigenetic modifications, particularly those that involve changes in postmitotic cellular genetic expression due to chromatin changes (the protein complex that surrounds the DNA). In this way, for example, neurons – mainly postmitotic – could have their function affected by agents that act by modifying the chromatin and DNA methylation and hence, altering cell function with no need that such change would be heritable. This amplification of the epigenetic concept becomes relevant in view that the environmental impact could act at different periods of life and also be potentially reversible, without requiring the heritability condition.

It is also interesting to note that recent advances on epigenetic processes classically limited to DNA markers have included changes on specific sites of the RNA (it transfers the DNA message to the protein synthesis process), which adds further complexity to these events linked with cell differentiation, regulatory in terms of epigenetics and transcriptomics[3,5] (see footnotes in sections 2.3 & 2.4) (Desrosiers et al., 1974; Jia et al., 2011; Luo et al., 2015; Wu et al., 2016; Dominissini et al., 2016; Mauer et al., 2017; Liu et al., 2016; see note on Willyard, 2017).

According to observations made by Cole (2009), socio-environmental processes regulate the gene expression in humans. In turn, they affect hormonal and neurotransmitter activity, which interact with cell receptors. These, in turn, activate genetic transcription factors. They also condition the fate of those genetic expressions that best adapt to conditions imposed by cultural demands. This plasticity stresses the tight relationship between genetic inheritance and socio-ecological mechanisms. In this regard, Jablonka and Lamb (2007) propose four types of heritability (evolution in four dimensions), they are: genetic, epigenetic, behavioral and symbolic, each providing variations over which natural selection would act. This bio-social interaction is of significative importance to understand a series of processes at the individual and collective levels, in the field of brain and behavioral organization.

Prudently, being of current knowledge in the field and the different susceptibility of genetic mechanisms between somatic and germ cells, Heard and Martienssen (2014) conclude reproducing a commentary made by S.J. Gould (1980):

> "In sharp opposition to biological evolution, human cultural evolution is of Lamarckian character. What we learned in one generation, we transmit it by teaching and in writing."

Within such a view, some factors could affect "fidelity" of transmission in this series of events. Thirty years have gone by since the clear thought by Gould (see later Meloni, 2014).

In order to approach an expanded view and potential implications of this sociobiological concept, it would be proper to mention that the various cultural conditions impose a differentiated context in which different genes would find the opportunity to express or not. This concept acquires particular significance to understand ethnic variations, but also includes a call to attention to the generators of public policies regarding the impact of the culture of poverty, indigence, and marginality – as much as the evolving *technocratic culture* – on individual and social construction.

Cultural inheritance constitutes an instrument of accumulated information that operates with a shorter time constant; hence more dynamic than biological inheritance (Tomasello, 1999). This allows to attempt an understanding of the acceleration of *H. sapiens'* abilities and strategies that seem to have taken place during the last tens of thousands of years, since language and graphism emerged and, with it, the progressive construction of a mind capable of generating symbolisms, abstractions, virtual structures, and values, and to trigger the development of accumulated knowledge. For this to happen, the presence of a brain organization was necessary to be capable of generating, storing, and processing such events, together with peripheral structures for its reception and expression.

Chapter 3

COMPARATIVE ISSUES

Subjacent to our mental constructions – whether virtual or symbolic – subsists our neurobiological matrix that processes necessary information on autoregulatory and survival modes. This tends to express itself through our overt or intentional behaviors – sometimes vetoed – and conditions our human relationships.

> "...our results suggest that there may be no fundamental discontinuities between social structure in humans and animals" (Chase et al., 2002).
>
> "Our brains did not grow to be four times larger until several million years after the lines split, in spite of the fact that our ancestors were bipedal 3.5 million years ago. This growth in brain size started and finished about 150,000 years ago – BEFORE the development of language, cooking, and agriculture. Everything that we tend to think of as "human" was produced by software programming (as opposed to hard-wiring) …" (Dennet, 2006).
>
> "Recent studies are consistent with the presence of general intelligence in mammals (rodents and primates) ...The presence of general

intelligence in nonhumans implies that modular[12] abilities can arise in two ways, primarily through automatic development with fixed content and secondarily through learning and automatization with more variable content. The currently best-supported model, for humans and non-human vertebrates alike, thus construes the mind as a mix of skills based on primary and secondary modules. The relative importance of these two components is expected to vary widely among species..." (Burkart et al., 2016).

Should we assume that we are biological individuals with social history and self-conscience, it would allow us to speculate with a greater realism that our overt behaviors have a probable origin and similitudes – essential drive components – with those of our predecessors, though optimized, camouflaged or replaced by the virtual and material mind and social structures that seem – or attempt – to fade out or vanish our true ancestral evolutionary origin. Such structures have led us to the development of excellence in several fields of human activity, as well as to individual or collective more questionable expressions.

Setting aside the sophisticated language development and the construction of virtual systems, if we review a series of behaviors from other species, we will find surprising similarities with some of our basic behaviors. Our evolution as a species, the development of new operational instances – intraspecies tactics and strategies – was made possible due to the reconfiguration of distributed, conditional neural nets. These were immersed in the notable cell density and connectivity increase of our neocortex and its subcortical (bottom-up and top-down) connections (see for example in Dunbar, 1993; Dunbar, 1999; Byrne, 2004; Striedter, 2006; Latzman et al., 2015). Also, evolution would have taken a ride on the reformulation of programs (based on the conditional connectivity of distributed neural ensembles) motorized by the demand imposed by the high degree of socialization of our species. That is, an extended ecological demand to incorporate the physical, emotional and cultural events of the

[12] Domain-specific adaptations or modules.

environment. Adaptation to such socialization and brain demand, of high energetic requirements of the human infant, highlight the biological and social importance of the extended period of the brain and mind postnatal development; even during the period of adolescence.

The growing sophistication of our cultural constructions and emergence of language as an instrument for cognitive and emotional construction allowed the consolidation of a brain and mind exercised in the dynamics of complex virtual social structures. It also continuously operates on brain organization based on neuroplasticity. The wonder that supposes the capacity of creating virtual worlds and of exercising complex behaviors involved in research, imagination, and creativeness based on the interaction between biological organisms built by connected universes of cells and molecules, represents a formidable event in the *Natural Kingdom*. Within this context, it is worth stressing the interrelated construction of cognitive and emotional dimensions.

3.1. The Brain under a Comparative View

It was previously mentioned the similarities between humans and non-human primates (especially with the chimpanzee; our living, genetically closer[13] evolutive congeners) concerning specific connectivities that associate brain structures with operative modules (Burkart et al., 2016). In a comparative methodological study (Rilling, 2014), images were taken using Structural Magnetic Resonance to compare brain proportions, Positron Emission Tomography (PET)[14] to compare glucose metabolism at rest, Functional Magnetic Resonance to compare neural connectivity, and

[13] Our common ancestor with the chimpanzee and bonobo – Genre Pan –, lived approximately 5-7 million years BCE.

[14] A procedure based on the injection of a radioactive tracer that is later detected in its localization and concentration. It is ocrrelated with the metabolism of glucose and blood flow, and taken as indicators of neural activity.

Diffusion Tractography[15] to compare fiber and white substance connectivity. These procedures helped to detect a more substantial presence of associative cerebral cortex in humans and to compare brain asymmetries (human lateralization of language and dexterity) that were initially presumed to be human, now found also in great apes.

Fiber projections (connectivity) and program plasticity mechanisms (software reprogramming) do not have equal distribution and comparative complexity. Hence, although there are several shared characteristics, there exist differences in the extension and connectivity of some brain regions, particularly in prefrontal, parietal and temporal association cortex, with fewer differences in sensory and motor areas (Avants et al., 2006). As mentioned before, dynamic resources – programmatic or software – would add a differentiated functional dimension with phylogenetically close species.

Collectively, observations tend to suggest that compared with other extant primates, there exists in humans more neural associated with cognitive processing rather than with perceptual processing. Additionally, the observation (Ghirlanda et al., 2017) that humans express a superior memory to remember and recognize sequences of events or stimuli would have allowed language development, planning, and complex cognitive processes. In this regard, the amount of time assigned to test completion does not seem to generate better performance in orangutans (Lind et al., 2017), thus reducing the possible impact of time constraints for a solution of the problem.

This greater neurocognitive complexity would not be germane to what Lind et al. (2013) called "cultural capacity", or generating complex cultural structures. These authors consider that its full emergence would not have occurred until more than 170,000 years ago, judged by a series of artefactual archaeological and biological (genetic) evidence, although of partial, primary existence previous to *Neanderthal* separation from the *H. Heidelbergensis* some 500,000 years ago. Such brain/mind potential ability

[15] This procedure is based on the detection of water molecules diffusion, and since they preferentially spread along axons, it allows to define neural tracts.

would have existed previous to its cultural expression, based on the fact that on its initial stages cultural evolution would have been extremely slow.

Another interesting phenomenon of brain dynamic that has been studied on a comparative basis is the so-called *default mode,* characterized by higher activity in specific brain areas during resting conditions than during attentional activity. This activity would be related to neural nets maintenance processes or introspective thinking. This default mode has also been described in chimpanzees (Barks et al., 2015) and Macaque monkeys (Hayden et al., 2009; Kojima et al., 2009). According to results obtained with PET, during such period there is an increase in blood flow and glucose consumption in prefrontal and medial parietal areas compared, for example, with the demand imposed by working memory. Hence, the possibility has been considered that similar processes of introspection or neural maintenance would also take place in these non-human species.

In several aspects, it seems that similarities and differences of our brains with those of great apes (chimpanzee in particular) are not limited to morphological aspects (relative size of some brain regions), but also functional ones. Schematically it could be said that, compared with the chimpanzee, in the human brain there is an optimization and relative increase in the complexity of neuronal circuits involved with cognitive performance and socio-cultural capacity (Lind et al., 2013).

3.2. Reward, Pleasure, and Risk-Taking

Be it of affective (emotional), monetary, physiological, social recognition, self-esteem or other, the concept of "reward" is deeply entrenched in our motivational mechanisms of behavior. From an instrumental point of view, reward has functioned as a behavioral inducer or reinforcement in all tested animal species, associated to the perception of pleasure – conceptualized as the resolution of tension due to a desired objective, or resolution of some as yet unsolved expectations (material, personal recognition, offering to an agent of virtual construction). This concept of pleasure is based on the resolution of a given expectative or desire, thus relaxing a state of tension (whether cognitive or emotional), which in turn generates a new (transient or enduring) emotional (affective) state.

Neural circuits related to reward and pleasure are interconnected. These circuits involve cortical brain regions (cingulate cortex, prefrontal areas), and also subcortical (ventral striatal and pallidal regions, septal areas, nucleus accumbens, thalamus), and mesocorticolimbic pathway (circuits that connect elements from the mesencephalon, cerebral cortex and limbic system), with a particular role of dopamine, according to Ikemoto and Panksepp (1999), Elliott et al. (2000), Knutson et al. (2001), McClure et al. (2004), and O'Doherty (2004). These components have a similar distribution in humans and non-human primates. Of course, brain organization supporting the complex mind virtual constructs (values, symbols, institutional concepts) and sophisticated language semantics establish functional differences related to behavioral (motivational) and cultural complexities. However, it also should be brought up clearly that the neurobiological bases involved with "reward" and "pleasure" recruit some similar components, thus sharing various circuits by humans and non-human primates (McClure et al., 2004; O'Doherty, 2004; Pessiglione et al., 2007; Berridge and Kriegelbach, 2008; Obayashi et al., 2009; Haber and Knutson, 2010; Schultz, 2010; Smith et al., 2011; Fumagalli, 2012; Azzi et al., 2012; Bissonette and Roesch, 2016; Sirigu and Duhamel, 2016; Tovote et al., 2016). Both, "reward" and "pleasure", are phenomenological pillars of animal behavior, including humans.

It is indeed interesting that mechanisms of defense and flight at the perception of risk, and responses to physical and emotional feeling of pain, and those associated with a threat, share a series of neural circuits placed in phylogenetically "older" brain structures, such as the mesencephalon, diencephalon and limbic system (archi- and paleo-cortex). This sharing tends to build up a common final pathway in the possible motor strategic responses (fight, flight, scream); that is, to establish a limited menu of possible behavioral answers. Thus, they represent evolutionary preserved neural circuits and behavioral schemes.

3.2.1. Pleasure and Risk-Taking

Pleasure understood as a resolution of a state of tension or displeasure has a fundamental role in behavioral construction. Searching for pleasure is intimately linked with an evaluation of risk-taking, a cognitive process involving a judgment concerning gain and loses – understanding this as a fraction or dimension of the non-achieved goal – and the opportunity in arriving at a decision. The higher the *profit*, the higher the risk to be taken and the pleasure of success. This "behavioral equation" involves inhibitory components of cognitive – evaluation of the risk taken – and emotional – the drive generated by the goal or reward sought – dimensions. These inhibitory mechanisms go through a developmental process with age, affected by the maturation of neural circuits – mostly prefrontal –, modulated by experience and education. Socialization – or covert domestication – impose the development of inhibitory processes as part of the intelligent, or adapted behavior. Personality and educational traits are involved in it.

Such inhibition is a behavioral survival characteristic at the individual level in all species. Inhibition at the social level, whether external or internal *veto*, is observed mainly in species that conform to gregarious communities. These respond to criteria imposed by the existence of hierarchies, that – depending on the particular species considered – regulate access to reproductive rights, feeding, consumption, earnings, knowledge, health, education, power. According to Sirigu and Duhamel (2016), and LeDoux (2012), in social behavior one of the central inhibitory mechanisms involves the dopaminergic system.

3.2.2. Deception and Reward

Several species – marine creatures, birds, and primates – use false leads to induce "cheating" or confuse their rivals, or reduce risks or threats. Deception in such cases has a survival value. Humans, too, express similar behavior, thus recognizing shared animal components. Though in the case of humans with their expanded culture and social structure, deception is used for other reasons too, namely, to obtain a gain, although it may not imply survival. Here is when Machiavellianism enters the scene, although deception and Machiavellianism are behaviors very close concerning the goal of survival or predominance. Machiavellianism has been characterized as a social strategy that involves manipulation of third parties in pursuing self-benefit, and often against the interest of them (Wilson et al., 1996, cf. Whiten and Byrne, 1997).

Gavrilets and Vose (2006) associate the development of Machiavellian behavior with social competition:

> "The "Machiavellian intelligence" hypothesis (or the "social brain" hypothesis) posits that large brains and distinctive cognitive abilities of humans have evolved via intense social competition in which social competitors developed increasingly sophisticated "Machiavellian" strategies as a means to achieve higher social and reproductive success" (Gavrilets and Vose, 2006).

From an evolutive and comparative point of view, Machiavellianism would stand for complex forms of manipulation and not for resources from a simple repertoire of deceptive nature, such as the song of the bird Tero respect to its nest site. Hence, it is required to differentiate the dimensions or complexities of the deceiving/Machiavellian behavior. For starters, it would imply a certain degree of social comprehension and reading of the possible mental state of the other(s), or to a potential hazard. Beyond the spectrum of possible complexities of the Machiavellian behavior, this involves a deceiving component, usually associated with the "social brain" (Sirigu and Duhamel, 2016; LeDoux, 2012). It is often expressed in societies that have reached a certain degree of complexity, like those living in large social groups, semi-permanent, as it is the case for carnivores, cetaceans, elephants.

The degree of socialization, and the demands that it implies for the development of effective strategies for social survival and reproductive success (would it be deception, manipulation, alliance formation, knowing weakness of others), has been related in *hominids* with tencephalic growth mostly of the neocortex (Dunbar 1998, 2009, 2011; Dunbar and Schultz, 2007; Reader and Laland, 2002; Byrne and Corp. 2004; Schultz and Dunbar, 2010; Boehm, 2012); especially in humans, in whom the brain represents approximately 2-3 percent of the body weight and in the adult in resting conditions requires 25-30 percent of the total metabolic expenditure. In this sense, what characteristics could our evolutive ancestor possibly have had?

According to some authors, the *ancestral Pan* lived in hierarchies of social dominance, which probably generated conflicts and coalitions in order to overcome them (Boehm, 2012; Byrne and Corp. 2004). In this sense, Dunbar (1998, 2009, 2010, 2011) supports the idea that brain size (particularly of the cerebral cortex) would correlate with the degree of socialization ("social hypothesis" of brain size) (see earlier). According to this, a complex deception ability (Machiavellianism) would also correlate in primates with brain size. This type of coalition is observed in gorillas (Robbins, 1999; Bradley et al., 2005), of which sexual favors – access to females – provided by the leader to potential competitors are occasionally used as a means to maintain a given hierarchical position in the social structure.

Besides, chimpanzees are skillful concealers, and in tests with humans, they hide food from humans and members of their kind (Byrne and Whiten, 1992; Hare et al., 2000, 2006; Wheeler et al., 2014). That is, certain behaviors aimed at the utilization of deceiving with self-benefit purposes are present in several species and are manifest in great apes and *H. sapiens*. These observations would suggest that the *ancestor Pan* would have practiced similar behaviors.

It seems evident that our brain and mind complexity has allowed us to excel in inventiveness, creativity, and the generation of beauty values and virtual constructions of a sophisticated degree that dominate essential dimensions of our behaviors. However, it seems also true that "below" such constructions, archi (old)-neural circuits or ensembles persist in our neurobiological matrix that find their expression under certain conditions and affect our human relationships. In essence, such basal circuits are involved with survival, regulatory mechanisms, flight/fight behaviors, territorial dominance, group prevalence, and modulated by paleo- and neocortical neural circuits. No matter how solid our cultural sophistication and "humanized" behaviors might be, the neurobiological scaffold of our basic behaviors keep their primal nature and are expressed under crisis conditions, or in programming to obtain maximal revenue from any possible transaction. Mainly when self-control relaxes, or circumstances reach critical condition, behaviors rooted in basic ancestral tendencies (drives for social privilege, prevalence, power or material accumulation) will emerge.

Reward and self-stimulation—an experimental approach

Experimental work in rodents by Olds and Milner (1954) demonstrated the existence of basal brain circuits (including structures such as the septum and hypothalamus) able to sustain self-stimulation. These authors implanted deep electrodes in rats' brains that would stimulate each time the animal pressed the feeding level to obtain food. They found that some basal regions would generate persistent stimulation (pressing the level although they would not eat the food pellets). This response would be related to complex processes such as motivation and reward, and opened-up the exploration of other brain regions; some involved in sensory systems (Jacques, 1979).

Intimately related to the animal behavioral matrix lies the notion of reward; be it for food, sex, a hierarchal promotion or a potential material gain. Consideration of the potential reward associated with a given objective activates a set of neural circuits. These are also present in humans and include specific brain regions/areas, both cortical and subcortical (n. accumbens, ventral striatum, and pallidum) in complex meso-limbic-cortical circuits (Rilling, 2014) (see also before). Cortical regions especially include prefrontal areas (ventromedial and dorsolateral).

In conclusion, the point to stress here is that those neural circuits, forming part of our decision-making mechanisms involving fundamental processes such as reward and motivation, do not substantially differ from those that are operative in non-human primates (Rilling, 2014). What may change are the events that trigger them since through our cultural mind-molding we have incorporated values, priorities and social norms. They respond to a limited group of primary drives (those connected with survival, prevalence, community or social bonding, child caring, reproductive success).

Although connectivity is only one of the parameters that participate in the functional construction of brain mechanisms (others include synaptic density, neurotransmitters, and receptors, ionic and molecular transporters), it clearly emerges that – with structural and dynamic variations – the neurobiological set-up recognizes a basic model with several components shared by other species, most clearly by other (non-human) primates. This issue will be further discussed in the following chapters.

Chapter 4

Identity and Moral Issues under a Neurobiological Approach

"One can take the ape out of the jungle, but not the jungle out of the ape. This applies also to us, bipedal apes" (de Waal, 2005).

"There are reasons to believe that affective experience may reflect a most primitive form of consciousness, which may have provided an evolutionary platform for the emergence of more complex layers of consciousness" (Panksepp, 2005).

"Animal research has shown that the ability to share and be affected by the emotional state of another is organized by basic systems: preoptic area of the thalamus, and paralimbic areas sub serving attachment-related processes, involving the brainstem. As a result of the importance of social connections for mammalian survival, these attachment systems appear to exploit the well-developed physical pain system, borrowing the aversive signals associated with pain to indicate when relationships are threatened" (Decety et al., 2012).

In order to analyze the possible mechanisms involved in the construction of complex human behaviors, the issues of identity, ethics, and emotions are briefly considered. The debate on these issues in the fields of philosophy, neuropsychology and cognitive psychology exceeds the limits of the present discussion and excuse the author from attempting their treatment here. The aim is to highlight evolutive bridges across behavioral issues regarding human (*Homo* in general) and non-human primates; that is, the animal component in our human nature. For such reason, it will be attempted to convene in concepts that allow their consideration and relatedness with such a goal.

According to what has been explored here so far, daily life behaviors – supposedly rational – are not devoid of emotional components associated with the affective dimension. Concerning social codes, they can generate adhesion or rejection/resistance, but in every case, they will involve neural circuits with components that recruit emotional states. Hence, it would be unreal to dissociate the construction and response to moral or ethical norms, from such components.

By *emotion*, we mean a mental state or overt behavior that involves a manifest alteration of visceral functions (heart rate, peristalsis, diuresis, mydriasis, perspiration, hyperglycemia) and psychological states (anxiety, uneasiness, guilt, irritability). The concept of *ethics* will be applied within the context of rules imposed by a given social organization on the behaviors of its members. In general, in gregarious communities, previous to the development of symbolic language, rules could have been considered as the product of a community's survival demands. A hierarchical structure, with relative dominance among its members' somewhat "institutionalized" such sets of rules, always conditional to a latent "tug of force" among them. In non-human species, its application is subjected to greater dependence on environmental (such as nutrient availability) and humoral (mostly hormonal) variables, the latter implied in the conditional leadership.

Identity will be considered as the set of behavioral pieces of evidence that contribute to a build-up a social profile of the individual. It could be envisioned that the construction of one own's narrative is a process that begins from proprioception when the individual recognizes itself as emitter and receptor of internal and external stimuli.

It is explicit that there is no intention here to enter a philosophical and historical dialectic on the origin of ethics and morality, but to gain consensus on specific basic concepts that could be applied and limited to the objectives of this essay. That is, regarding a species comparative analysis and the possible neurobiological substrate.

In behavioral construction, the role of neural circuits in humans recognize a significant analogy with those involved in similar behaviors in non-human primates organized in communities with hierarchical structures, as discussed earlier. That is, the organization of complex behaviors in our species recognizes an evolutionary conserved primary neurobiological substrate this is present in other animal species. Hence, the physiological mechanisms subjacent to individual and collective behaviors recognize common elements and processes regarding programmed and emotional components (Fumagalli and Priori, 2012; Pascual et al., 2013).

Given the complex nature of the concept of morality, the neurobiological substrate has characteristics of distributed neuronal ensembles with nodes that would depend on the predominant component, whether cognitive or emotional. Among them, responses with manifest emotional components involve circuits located in the ventromedial and orbitofrontal prefrontal cortices, and parietal and temporal limbic system circuits. Dorsolateral prefrontal circuits would be involved in more "utilitarian" responses or associated with *responsibility judgments*, in association with circuits placed in the anterior cingulate cortex and striatal (subcortical) structures.

Judging from comparative studies (Preston, 2013; Decety et al., 2012; Santos and Rosati, 2015) regarding the construction of manifest behaviors in other species, it seems worth stressing the participation of similar neural nets associated with primal neurobiological substrates. They would be mainly involved with survival, constant evaluation of gain/loss (error) concerning the expected result, and the relative amount of "pleasure" (affecting drive) generated from such an evaluation.

From a comparative view, this is expressed in chimpanzee communities, the extant non-human primate with the most similarities with the human. In particular, this relates to emotional components in decisions that are supposedly rational (Santos and Rosati, 2015), and the genome, as mentioned.

Several non-human animal species recur to similar strategies in order to resolve or anticipate potential conflicts; reciprocity, mediation, empathy, political agreement. In most species, such conflicts usually arise due to primary competition on territoriality, feeding resources, reproductive advantages, and social prevalence. These "competitive arenas" are evolutionarily conserved, though in humans they have been expanded and "camouflaged" under virtual constructs such as values, beliefs, or ideology. According to Preston (2013) and Decety et al. (2012), they represent the primary links in the formulation of ethical constructions. After all, several human values are built on how we offer a solution to basic conflicts between individual drives and community demands. These are associated with material possessions, revenues, privileges, self-esteem, and social prevalence. All of them are not germane to the spectrum of possible behavioral motivation observed in non-human primates, except for creativity, accumulation of knowledge which is an "altruistic" behavior concerning non-related individuals, and a sense of beauty – all seem to be *human "markers"*. However, there is an additional "marker" of humanness, that is, its ability to act cruelly; to inflict avoidable suffering, or inflict pain to third parties with no other cause than personal desire or to obtain an advantage. Such aberrant behavior belongs to psychopathology only seen in the human domain, whether as an individual or institutionalized behavior.

The fact that language and its complex structural developments allowed to expand to the virtual dimension our cultural universe does not displace the motivation that generated the mentioned primary – ancestral – drives, on which we construct our social interaction. That is, at the very base of the "pyramid of most human constructs" are remnants of old, primal, behavioral inertias.

One of the nodal points associated with motivation and reinforcement, as well as assigning a value to given signals, would include neural circuits in the orbitofrontal region of the prefrontal cortex (Azzi et al., 2012), with its activity and connectivity modulated by the social (environmental) context. Lesioning this cortical region in human and non-human primates results in changes in personality and social interaction, with loss of fear and aggressiveness. According to these authors, this would be one of the circuital components involved in behavioral dynamics associated with moral issues, together with those involving the ventromedial and dorsolateral prefrontal cortices. Additionally, the cingulate cortex would include neural components entailed in detecting judgment errors in utilitarian responses.

The fact that neural circuits involved in attributing value to given behaviors would also be part of other behaviors (cognitive and emotional) suggests that there is no "specific" brain area related to the *moral brain*. The latter would recruit neural circuits from several cortical areas – involving both cognitive and emotional responses – that participate in the construction of other behaviors (Moll et al., 2002; Pascual et al., 2013). This would support the concept of "an organization of conditional and distributed neural ensembles" involved in moral conducts aimed at the detection of desirable or acceptable ones. In this respect, as indicated by Latzman et al. (2015), gregarious groups manifest basic social constructs involved in the detection of acceptable and unacceptable conducts for any given hierarchical social organization. These basic, implicit codes are construed in humans as moral and ethical rules that guide desirable and punishable behaviors.

In non-human primates, cooperative behavior – the basis of community ethics – is typically expressed among related individuals with whom there is usually reciprocity, and is rarely extended to non-related individuals (Silk et al., 2005). This behavior takes place in gregarious species, particularly with individuals from the social environment, as an evolutionarily conserved phenomenon of empathy typical among humans (Silk and House, 2011; Decety et al., 2016). Several factors influence empathy, most notably sex hormones such as testosterone[16] and

[16] Crucial for the expression of a series of events that conform to male phenotype, particularly specific social behaviors as aggression, territoriality, reproductive drive. Androgenic and estrogenic receptors mediate its effects following its aromatization to oestradiol 17β (see Karlsson et al., 2016).

neurohormones as oxytocin – a peptide[17] released by neurons located in the hypothalamus, at the base of the brain. According to De Dreu et al. (2011), this peptide affects the parental care of the offspring and would be involved in the mechanisms of intragroup acceptance and intergroup rejection, both of fundamental significance in prosocial behavior and social organization.

Its role in offspring care involve neuronal circuits highly conserved in all mammalian species; particularly those neurohormonal mechanisms that involve the autonomous nervous system and neuroendocrine system, the brain stem, preoptic area and basal ganglia (Panksepp, 1998; cf. Decety 2011; Decety et al., 2012; Panksepp, 2005). According to Decety et al. (2012), empathy among humans is built on primary forms of consideration to third parties, involving subcortical and neurohormonal mechanisms associated with social links and parental care. This is also expressed in non-human primates, although usually not involving socially unrelated individuals.

This evolutive mainframe of empathy is also affected by social preferences, which complicate the application of fair decisions (ethical framework). According to Montoya (2013), such judgments involve cognitive and emotional components and is sensible to testosterone and estradiol levels during the process of early hormonal imprinting, with the prevalence of testosterone inducing a utilitarian moral criterium according to Chen et al. (2016).

It should be pointed out that in humans, the more complex brain functions (of cognitive and emotional nature) took place without displacing more ancestral phylogenetic levels. In this sense, according to Decety et al. (2012), brain organization operates from multiple entries processed in parallel so that responses will depend on the interaction of those levels.

From a historical point of view with an evolutionary perspective, let us mention the "superposition" of levels in phylogenetic development as proposed by Christofred Jakob (1941) ("archi-, paleo-, neo- neuronal

---

[17] Molecule formed by a series of amino acids.

systems")[18], anticipating the concept of the "triune brain" proposed by Paul MacLean (1990).

This fundamental notion of "interactive aggregation" of neural circuits or "ensembles" in the evolution of brain organization represents perhaps the clearest "biological scheme" regarding the emergence of the "visceral brain" in humans[19,20]. It also illustrates that the "basic" (primal) programming (drives, regulation of the internal milieu) and more complex executive levels are integrated into behavioral construction, and are not uniquely dependent on neocortical processes. As pointed out by Decety and Cowell (2014), subcortical circuits are not systematically subjected to a top-down interaction involving neocortical neural assemblies, which probably are accessed in order to decide behaviors – whether manifest or introspective – under conditions of high ambiguity and low predictability (Parvizi, 2009).

An emotional and cognitive reaction to suffering would depend on the relatedness or social proximity, and to cultural patterns of different ethnic groups. According to Lamm et al. (2011) and Decety and Cowell (2014) in non-human animals, these emotions are associated with the pain circuit – involving the anterior cingulate cortex, anterior insula, supplementary motor area, and periaqueductal grey substance. This circuit ensures high efficiency and an evolutionarily conserved final common pathway.

In summary, although the concept of emotion – and its evaluation – still generates some discussion (see LeDoux, 2012), the neural circuits that take part in the parental response to signals of suffering and vulnerability

---

[18] The chronological development of the concept of the "visceral brain" involves the germinal participation of Christofried Jakob (1866-1956), a German-born scientist settled in Argentina, where he contributed to developing brain research and the neuropathological school (see Triarhou and del Cerro, 2006; Triarhou, 2008).

[19] It corresponds to a mention here on the influence of the Pavlovian neuroscientific Russian school of the XX century, and its contribution to the concept of cortical-visceral *temporary links*.

[20] An experimental demonstration of the *visceral brains*, as further developed later by Bykov (1958) and Bykov et al. (1959).

from the offspring and related individuals involve neuroendocrine[21] and autonomic systems and neural circuits evolutionary conserved among mammals. The subsequent "disposition" or "indifference" to react to signals of suffering or vulnerability has been incorporated to value judgments – moral –, thus reflecting the intimate biosocial relationship, in this case involving neural and humoral responses.

---

[21] Interestingly, in the case of oxytocin, besides acting postdelivery, inducing parental care of the offspring and attachment in adult humans instead of inducing prosocial behaviors, it would reinforce the evolutionary trend of discriminating among within- or inter-group individuals, thus increasing the expression of ethnocentrism and eventually xenophobia (De Dreu et al., 2011).

Chapter 5

SOCIAL ECOLOGY AND BRAIN REORGANIZATION

> "Panksepp (1998) provided evidence that the social attachment system is built up from more primitive regulation systems such as those involved in place attachment, thermoregulation, and physical pain" (Decety et al., 2012).

In previous chapters, several comparative aspects were reviewed regarding our evolutive background related to our general behavior as a species. Based on a geological metaphor, our basic neurobiological, and cultural background and dynamics, could be represented as *geological tectonic plates* with their potential *friction* as a result of their interaction capable of generating mutual changes and productive behavioral events. That is, on one side, brain/mental effects are due to continuous adjustments of our *primal animal neurobiological expression* of drives and basic needs being modulated by, or contingent to our evolving cultural dynamics and demands. The cultural context and individual behavior are in turn affected by neurocognitive and emotional profiles that result from the expression of our primal neurobiological scaffolding.

Metaphorically, the bio-social interaction could be described by the following geological sequence of events[22]:

> "Water is stored in the crystalline structure of minerals, Shillington wrote. The liquid gets incorporated into the Earth's crust both when brand-new, piping-hot oceanic plates form and when the same plates bend and crack as they grind under their neighbors. This latter process, called subduction, is the only way water penetrates deep into the crust and mantle, but little is known about how much water moves during the process…"

Both dimensions of our basic construction of heritable behaviors have been depicted as "tectonic plates" in order to underline the continuous "tension/friction" between them resulting in the various individual and collective behaviors; whether constructive, creative or destructive. Personality traits will add an identity color and behavioral trends to the product of such *friction*. Certain domains of the cultural "plaque" will generate either softer or harsher consequences, as they may confront with primal behavioral drives with positive evolutive pressure such as survival, group hierarchy, and prevalence. These are represented by conditions such as inequity, poverty, indigence, and marginality.

A demonstrative example of the interaction between the socio-ecological environment, brain organization, and social behavior is offered by studies performed in coral reef fishes (Robertson, 1972; Maruska et al., 2013). In these species, the interaction between social group composition and neurobiological-neuroendocrine events can be experimentally analyzed. For example, certain coral fish communities are characterized by a dominant male followed hierarchically by a cohort of subdominant females. Should the male disappear, one of the subdominant females will undergo sex reversal and become a dominant male. These observations serve as an example that each species will express in its fashion the impact

[22] https://www.livescience.com/64091-earth-is-eating-its-oceans.html?utm_source=lsnewsletter&utm_medium=email&utm_campaign=20181118-ls (based on *Nature* 563 (2018): 389-392.).

of changes in social ecology, according to their genetic and functional characteristics. Humans, as a hyper social species immersed in its cultural dynamics, will express subtle or manifest biological adaptations to the ever-changing cultural ecology and social structure, depending on the character and identity construction of each of its members.

The mentioned case of the coral reef fish has allowed for further studies at the genetic and molecular level of the dynamic interaction between the social environment and biological configuration (O'Connell and Hofmann, 2011; Oliveira, 2012; Maruska et al., 2013).

As said, individual construction is based on two interactive vectors; the genetic and the ecological (social and environmental). Concerning the analysis of the basic human construction, characterizing the behavioral profile of our common ancestor as expressed by bonobos and chimpanzees, leads to two different characteristics regarding social organization and ways to solve disputes. Whether empathy predominates as in bonobos, or reciprocity in chimpanzees (de Waal, 2005), the former is more prone to search for consensus and female predominance in case of litigation, while the latter is more belligerent and under male predominance. According to authors such as Furuichi (2012), *Pan troglodytes* express a greater tendency toward aggressiveness among groups and the generation of male coalitions, while *Pan paniscus* include females in their coalitions which tend to manifest more pacific behaviors to resolve their disputes. In the latter, no lethal aggressions have been described among its members (Gruber and Clay, 2016) and show a lesser tendency to form strictly male coalitions (Surbeck et al., 2000). These profiles are affected by environmental conditions (Surbeck et al., 2017) and the existence of subspecies (Gruber and Clay, 2016).

Both – bonobos and chimpanzees – probably reflect primal characteristics in the population of the *ancestral Pan*, suggesting the possibility that the human genome could be a carrier of both tendencies expressed with varying strength in the various human populations, depending on the conditions provided by the social ecology. Hence, it would be attainable to believe that our species might represent a mosaic of such trends that are modulated by cultural constructions, perhaps reflected in our various ideological and social constructs and the condition of tension/relaxation among different states of ethnic and social groups. In this context, Wrangham and Glowacki (2012) ask if humans developed specific behavioral adaptations to conflicts and propose that the "chimpanzee model" is apt for the analysis of the biological and cultural evolution of belligerence in humans. The concept of "human bipolar nature" has been proposed by Boehm (2012), according to whom *H. sapiens* would retain and express both behavioral trends.

The subsequent development of written and oral language by *H. sapiens* allowed the incorporation of a series of virtual constructions, symbolism, and forms of communication, allowing for a superstructural cultural construct "above" our ancestral animal condition. Among other consequences, it allowed humans to navigate on different cultural domains, generating the sensation that we departed from our ancient roots. This seems to be true as long as we maintain ourselves diving in that new – virtual and material – dimension of human creativeness. However, such development did not erase our ancestral neurobiological construction. It generated a somewhat "conceptual schizotypy" where "ancient ghosts" of dominance, territoriality, reproductive rights, feeding rights, anger, cruelty, and power seeking, merge whenever we abandon the attempted shield of our "new" human, superstructural cultural dimension – a dimension fed by creativeness and knowledge-seeking drives. In that ancestral construction there underlie sophisticated socio-cultural developments of our species and political behaviors. After all, what is politics about if not the search for the build-up of consensus or strategies with which to manage behaviors associated with common needs, privileges, human rights, group prevalence, territorial protection or expansion, or material accumulation? They seem to represent old vices under new disguises.

Several other studies followed the experimental studies by a series of researchers, such as Harlow (1959) and the groups of Greenough and Rosenzweig (Krech et al., 1960; Volkmar and Greenough, 1972; Turner and Greenough, 1985). They confirm that the physical and social conditions of the environment during the breeding period and later affect brain development. This takes place at the microstructural and neurochemical level; a cognition influences cognitive abilities and emotional behavior in adult life.

In this realm, neuroplasticity (see before) provides an opportunity for adaptive and maladaptive processes, depending on environmental cues. This "valued neurobiological money"– ncuroplasticity – offers grounds for positive and negative outcomes (see a revision in Fauth and Tetzlaff, 2016). It could result in the optimization of brain/mind resources, or its impoverishment, depending on cognitive and emotional contention and enrichment of the environment where it grows.

In other opportunities, it has been commented that every educational process implies the installation of a given mind reconfiguration based on dynamic brain connectivity of distributed neuronal-glial ensembles. Hence, environmental cognitive and emotional enrichment, and impoverishment will have significant consequences on individual performance and development. In this context, public policies when based on exclusively financial factors of the complex social equation will result in condemning vast public sectors to environmental conditions that negate and suffocate human potential. Under this condition, it becomes a moral problem – a cruel picture of human degradation – and a source of ballast for social development.

Chapter 6

Evolutive Background of Poverty and Social Marginality: Poverty and Marginality as Social and Economic Categories

"The predatory phase of culture is attained only when the predatory attitude has become the habitual and accredited spiritual attitude for the members of the group; when the fight has become the dominant note in the current theory of life; when the common-sense appreciation of men and things has come to be an appreciation with a view to combat" (Thorstein Veblen, 1899).

"Social status is one of the most important predictors of the quality of an individual's social environment. Consequently, social status has been intensively studied in non-human primates, including as a model for the effects of social stress and socioeconomic status (SES) in humans. Social status in non-human primates is encoded by dominance rank, which defines which individuals yield to other individuals during competitive encounters. In settings in which hierarchies are strong, enforced or subordinates have little social support, low dominance rank can lead to chronic stress, immune compromise, and reproductive dysregulation" (Tunga et al., 2012).

Basic tenets within the animal kingdom involve securing enough territory for purposes of feeding and reproduction, which results in intraspecific belligerence due to competition. In gregarious animal communities with a hierarchical organization, the extra personal space and social rank – with its implied privileges – is subjected to similar behavior. *H. sapiens* has transformed its intraspecific dominance into a predatory culture applied to the opponent or "different" – whether of social, ideological, financial or religious nature – that is aimed at eliminating (or absorbing) the competence. This belligerent attitude incorporated to the cultural construction may acquire characteristics of open or covert strategies; in the latter case aimed at suffocating the capacity for informed criticism of large human groups using different resources (poverty, indigence, marginality, promotion of cultural banalization, and consumerism as a means to generate self-satisfaction). The *predatory culture* among humans acquires different *costumes* depending on whether it takes place during war or peacetime condition. The goal is to subdue actual or potential opponents by either physical (war, economic/financial pressure) or psychological means. The latter case is accomplished by generating emotional (beliefs) dependency, and cognitively dormant or indifferent communities and individuals that are displaced toward the marginality of critical knowledge. This allows for programmable domestication or undermining their potential resistance into intellectual submission. This is one reason the promoted term *globalization* represents a limited communicational metaphor restricted to certain socio-economic strata or dealing with profitable operations – information regarding critical events and long-term policies with public incidence –, i.e., management of geopolitical issues, corporative revenues, financial decisions, and is accessible to those operating within specific socio-economical strata. For such reasons, access to information having a public impact only reaches privileged positions as one is closer to higher institutional levels or ranks on educational background, but by no means such – critical – information is functionally globalized. Instead, it is functionally "sequestered" (hard to have access to it, or to a critical, informed, view) from public knowledge. Granted, specific knowledge of the different human cognitive domains by no means have public accessibility. Hence, there is a need to culturally regenerate the behavior of public policy makers and leaders to not profit from such differential, privileged access to critical information. Also, public media must increase public awareness of cultural trends and scientific-technological developments with potential social impact.

In gregarious communities with hierarchical organization, access to nutrient resources – more yet during periods of low accessibility – is decided on a hierarchical arrangement of rights which eventually could result in some form of alliances or concessions, or, inhibitory/repressive commands. Hence, in this type of communities, the hierarchical structure decides the possible levels of survival and welfare of its constituents, depending on climate conditions, food supply, hormonally-conditioned leadership and the individualized genetic imprint (Kerhoas et al., 2014). This social structure could result in the competitiveness for food, reproductive rights or hierarchical prevalence, and potential confrontations could be terminal.

Once again, dominance – whether territorial, or economic or political reflected in the intragroup positioning – is a determinant factor in the individual and collective behavior in gregarious species with hierarchical structures – based on force, or delegated institutional rights as in human societies. In our *sapiens* culture, access to strategic minerals, drinking water, energy sources, communicational technology, and fertile lands promotes belligerence or terminal wars. In such a process, social and economically low ranked citizens found themselves marginated from most aspects of the industrialized world (see ahead), or from meaningful information, and tend to become instrumental to hierarchical decisions as labor- or fighting forces.

In the concept of Richard Hawkins (1989) relative to "the selfish gene", the multiple morphological and functional expressions of living creatures respond to a single goal: genetic survival. On that basis, the dynamics of non-human animal gregarious groups with hierarchical organization result in the survival of the fittest and priorities down the hierarchical levels, that favor genetic or phenotypic expressions (see ahead). Since cultural expressions also possess a parochial and hereditarian component, the defense of such cultural heritage has the potential of generating similar confrontations with expressions of dominance, predation or annihilation. In our species, the genetic expression is altered by the emergence of cultural values and priorities (see before, *biosocial interaction*), due to ethnic characteristics or socioeconomic conditions, manipulated by institutionalized structures and relative dominance.

In our times, the construction of the concept of "property" and "privilege" seems to emerge from those primary tendencies mentioned above, expressed in non-human animal communities with a gregarious organization and hierarchical structure. By chance, do they represent an evolutive primal background of the sophisticated socio-cultural construction that *H. sapiens* then undertook? Does it represent an equivalent of the conditions that generate poverty and marginalization in societies along the history of human civilization? Do poverty and social marginality in our human societies represent a culturally reformulated or derivated form of community behaviors evolutionarily conserved in gregarious species with hierarchical structures?

The ancestral interspecific competitiveness and intraspecific cooperation have undergone a reconfiguration to generate leads and privileged positions relative to social consideration, access to power and dominant positions. In this intraspecific competition, factions have frequently acquired belligerent profiles of religious, financial, ideologic or cultural fundamentalisms which have the common characteristic of social or cultural predatory behavior. Whatever the intellectual disguise used, the common denominator is the fight – competition – for power or a position of dominance or prevalence. Its goal is achieving enhanced power based on attracting trained human minds, or extraterrestrial mining, or communication prevalence – a bloodless, surreptitious, highly efficient instrument among competitors. Emotional dependence and Machiavellianism have been productive strategies. However, in such struggle, where is the prosocial species character of our human nature?

One individual and collective behavior that appears to have been evolutionarily conserved has been that of accumulation. In nature, this behavior represents an adaptive one, due to seasonal changes in food access. In human societies concerning social psychology, besides periods of crisis or wars, accumulation – physical goods or financial assets – could represent an "evolutive dystopic behavior" tractioned by ancestral drives, but incorporating components of dominance, self-assuredness, whether individual or group – corporative.

Within the above context, it would deem necessary to place some considerations on comparative aspects regarding the generation of social marginals in non-human animal gregarious societies with a hierarchical structure – perhaps a parameter of the concept of poverty applied to contexts within the *Natural Kingdom,* that imposes "rights" or "privileges", on feeding and reproduction. Those derived, evolutionarily conserved behaviors would represent a continuum in human societies with constructs such as "poverty", "indigence" and "marginality". Hence the inquiry, is "poverty" an economic category exclusive to human civilization dynamics, or does it recognize similar expressions in natural history?

6.1. A Comparative Concept of *Poverty*

Let us remember that the number of individuals that composed an early *sapiens* troop or – later – a community was of fundamental importance since it implied a continuous adaptation of the social structure and interpersonal relationships to food availability. The number increased with the advent of a settlement, the productive rendering of the earth and the emergence of language. In such social development, in a gregarious community with a hierarchical structure in a species as ours, there emerged the need to establish complex rights and priorities that derived into norms; see the institutionalized "fictions" mentioned by Harari (2015). This development resulted in further arrangements to distribute power, territory, nutrients, and reproductive rights within the community which deepened the formation of social strata, with comparative differences in the opportunities for individual growth and social ranking. Progressively, this resulted in further segmentation concerning access to those requirements, as well a role in the production chain, access to information, and a share of the produce.

Figure 4. "*With God on our side...*" (Bob Dylan) (Modified from es.wikipedia.org.) "Dresden Destroyed. Scene from City Hall, 1945." The author is R. Peter.

One question to pose in this context is whether relegated members of the troops of *Homo* or anthropoid primates or, later on, of the primal *Homo* communities represented an archaic version of the present day socio-economic structure and marginality. This would be a forewarning of what several centuries later would become "poor or dispossessed", "destitute" and "marginals" in modern human societies. These conditions are not unusual in gregarious animal communities with hierarchical structures. Therefore, there are grounds to believe that under such social organization, from a comparatively evolutionary-conserved view, it would project onto human social structures.

"Social mobility" would depend on individual cognitive and physical conditions as well as in the ability to construct socio-political arrangements to acquire specific rights. These factors appear functional among several species, as it has been described for gorillas. At the human level, those factors appear affected by extrapersonal conditions since, after all, equal rights and equity do represent still unsolved issues in our modern societies. Let us consider that, as mentioned by former UN General Secretary Ban Ki-moon (2015), 2.5 billion people in the world lack adequate sanitary services and warned that, in one decade, 1.8 billion people would suffer from drinking water shortages.

Along with natural history and still ongoing under our human conditions, this somewhat "social Darwinism" would have been interacting with biological Darwinism, resulting in the expulsion or annihilation of displaced members. This type of "social eugenics" would have set up the basis for political – institutional – initiatives with similar objectives, although more complex and systematic in its makings and forms.

New social categories appeared in the community framework with the advent of farming settlements and the exploitation control of the land yields becoming a source of economic and political power and a private target. The notion of "territory" acquired value as an economic asset.

This issue merits a specific multidisciplinary treatment by itself and, in such sense, it is interesting the conceptual development of "poverty" as an economic category (Emilio Colombo, 2007). Kohler et al. (2017) dealt with the disparity development of richness since the end of the *Neolithic* period in Eurasia, associated with land property and increased farming and domestication of livestock. The concept of territorial dominance exerted by animal species probably has represented an evolutionary inheritance that primed human history of "private property". Hierarchical social structures would have promoted this outcome. The emergence of the economic concept of private property as a cultural construct probably added a new element in the construction of hierarchical societies and social inequalities.

Chapter 7

EMOTIONAL COMPONENTS IN THE CONSTRUCTION OF BELIEFS (I)

From Animism to Dominance[23] *and Psychological Dependence - Conditions for the Emergence of an Instrument for Individual and Collective Emotional Influence - Magical Thinking and Religiosity*[24]

The following concepts of Frans de Waal (2005) seem most opportune to be quoted here:

> "Modern religions are only a few thousand years old. It's hard to imagine that human psychology was radically different before religions arose. It's not that religion and culture don't have a role to play, but the building blocks of morality clearly predate humanity. We recognize them in our primate relatives, with empathy being most conspicuous in the bonobo and reciprocity in the chimpanzee. Moral rules tell us when and

---

[23] For ethology, dominance represents the possession of a high social status in a community with a hierarchical social organization.

[24] *Religion* (Shorter Oxford Dictionary): "Action or conduct indicating a belief in, reverence for, and desire to please, a divine ruling power...Recognition on the art of men of some higher unseen or power as having control of his destiny and as being entitled to obedience, reverence, and worship" (cf. Renfrew, 1994).

how to apply these tendencies, but the tendencies themselves have been in the works for time immemorial."

The consideration of the emergence of beliefs and rituals involved entering into an emotional realm, in this case, related with evolutive and comparative events, since the ritualistic type of behaviors has been described in anthropoid species. Its consideration may prove of interest in terms of analyzing human individual and collective behaviors.

7.1. General Approach

"Fear of the invisible is the natural seed for what each one of us considers religion. Thomas Hobbes, Leviatán (1651)" (cf. Sagan, 1995).

"...the core of all religions can be seen as an evolved mental adaptation to cope with the bewildering magnitude of phenomena which plague the exceptionally large human cortex." "The early human mind, this is the position taken in this contribution, needed a powerful paradigm which was able to explain all the seemingly unconnected and unpredictable events of nature as well as its awe-inspiring beauty and functionality and to make sense of the complexity of human wants and actions. Additionally, our ancestors needed rituals to interact with the supernatural powers who were believed to have created and to be in control of the world. Helplessness, stress, and anxiety were reduced through belief and ritual. It is no wonder that we are, by our very nature, *Homo religious*" (Schiefenhövel, 2009).

As stated earlier, one sensible point to stress regarding the final statement of Schiefenhövel (2009) is not that we "are" – which might inspire the concept of an intrinsic characteristic of our species –, but that we "became" *Homo religious.* Despite the emotional factors from which would have evolved such a condition and resulted in an early "imprinting" of the *Homo* brain – a somewhat "brain incarnation of religiosity or dualism" –, later cultural interaction and neuroplasticity provided grounds to "shake off" some of those emotional ballasts. Besides its potential anxiety and stress reduction effects, they do not form part of the basic conditions for physical – genetic – survival in the animal kingdom (*territory* where to exert *prevalence* on *reproductive rights* and access to *food resources*). Rather, a progressive cultural construction was demanded in order to cope with increased perceptions of physical insecurity and transience – an emotional challenge that grew with *H. sapiens´* cultural development.

The generalized presence of beliefs and religious constructions in *H. sapiens* suggest that the building blocks have a deep, ancestral origin. It could be discussed in which evolutive period it would have had its origin. However, at some point in the evolution of the *hominins* and *hominids* (if including great apes) the self-consciousness of fragility and geo-climatic dependence – whether threatening or propitiators of resources – would have ensued. These contributed to the progressive construction of an emotional disposition to attach to natural phenomena virtual – agency – properties, as well as behaviors – intentionality – that could emotionally shield them from such threats. One accessible way to deal with them has been attributing characteristics of agents with whom some form of interaction could be attempted. Virtuality entered the scene. Also, so did dualism – attribution of agency and intentionality to material and virtual beings and our human construction.

According to Panksepp (2005), the emotional experience would represent the most primitive form of self-consciousness which has served as a platform for more complex forms of consciousness. It would be opportune to add to this concept the notion that identity construction depends on our own and exteroceptive afferences, integrated and stored in our mnemonic records. That is, the construction of our narrative, or identity, is a process that begins from the proprioception when we recognize ourselves as emitters and receptors of internal and external stimuli/signals.

According to Peoples et al. (2016), animism would represent the "founding stone" for the posterior construction of beliefs associated with agents from the "ultra-world". Imagination, a building block of creativeness was inserted on the evolutive road of *Homo*. As defined by Tylor (cf. Peoples et al., 2016):

> "Tylor defined animism as a general belief in the "animation of all nature" (1871:258) and fundamental to religion. Animism includes a "belief in personal souls" (1871:260) as well as "a sense of spiritual beings inhabiting trees and rocks and waterfalls" (1871:260). We define animism as the belief that all "natural" things, such as plants, animals, and even such phenomena as thunder, have intentionality (or a vital force) and can have an influence on human lives."

Accordingly, the emotional components would have dominated the progressively more complex cultural construction. These were related to ecological interaction as well as among members of the community concerning identity expression and hierarchies.

As it will be discussed later, the emergence of a complex cultural construction has opened a debate of whether it represents a progressive process or a "revolutionary" one – as a quasi-discrete temporal event – of the last 40,000-50,000 years.

The following comments do not pretend to question the values of different beliefs or religions; values that could also be proposed from a non-religious or even agnostic point of view. The goal is to explore the possible evolutive and neurocognitive processes that may be subjacent to the construction of virtual, fantastic, worlds that would reside in some out-of-human-reach dimension. This includes the attribution of intentionality to beings, environmental objects and events of geo-climatic phenomena that would affect individual or collective "destinies", and to consider whether such construction recognizes previous manifestations in comparative anthropology.

7.2. Cultural Development and the Emergence of Virtual Constructions

Mithen (1996), a supporter of the idea of *the human revolution*, underlines two crucial periods in cultural development:

> "The two really dramatic transformations in human behavior occurred long after the modern size of the brain had evolved. They are both associated exclusively with *Homo sapiens sapiens*. The first was a cultural explosion between 60,000 and 30,000 years ago when the first art, complex technology, and religion appeared. The second was the rise of farming 10,000 years ago" (Mithen, 1996).

As mentioned previously, this interpretation of human cultural development as a "revolutionary" – or accelerated – process has been questioned by authors that stress evidence accumulated in times previous to the last 40,000 years, to which they suggest a progressive development (*"The revolution that wasn't")* (McBrearty and Brooks, 2000; see also Churchill and Smith, 2000). On speculative grounds, it is possible that once a certain cultural complexity has been met – a *cultural threshold* –, changes would have acquired a new dynamic – also promoted by the language acquisition and symbolic expression –, as we can see in our days with the dynamics of scientific-technological development. The concept of a long, progressive development contains several events that support it; for example, the artifacts – instruments; single-use decorative, other shells collars and dyes – that suggest expression of the identity in the egocentric space.

In this context, the basic concept of a culture related to objects, integrated with mental processes of growing complexity, led Hodder (2014) to propose the entanglement of human minds with objects of effective use that became symbols in the mental and social construction, generating a growing dependence with them. This concept of *mental incarnation* generated by human's object creations applies not only to the technological domain but also to ritualistic constructs as well as their symbolic representations (of gods, icons, amulets, talismans) and in our modern world to also goods that support our life comforts.

Occasionally, humans attribute intentionality to some of these objects from different origins and forms – talismans, amulets –, or virtual representations, generating some magical dependence with them. Perhaps the more ancient and persistent "entanglement" of humans is with their fears, probable generators of animism and ritualistic behaviors. In ancient times, and present days also in certain traditional cultures, such entanglement was limited to the search for protection or security, or providing resources for the community members against hard times, or the fearsome unknown.

In the non-human, animal universe, the reaction against a possible undisclosed threat – e.g., climate, fire, potential predator – is flight, hide or behavioral freezing up. However, how could one proceed with self-created threats? On similar grounds, there is no accessible information among nonhuman animals depicting a case in which a specifically designed object is interposed to attempt to interact with forces of unknown nature; that is, virtual or material events to which intentionality is attributed, seeking physical protection or reduction of the emotional impact of such a threat.

Quite to the contrary, human imagination has developed all types of instruments, whether material or virtual and behaviors attempting to reduce fear, anxiety or stress. Projecting intentionality to objects or phenomena and rituals were expected to install some form of interaction with unknown powers and attributing an "identity" to natural events, whether climatic, animals, unusual rock forms, generated an imaginary "channel" to interact with such agents. These were incorporated into their familiar universe or tuned them "accessible" to present their prays and alms.

Therefore, beliefs would have begun by assigning an activity or "intentionality" or agency to objects, animals and natural phenomena, or representations of them. In this context, primal parietal animal drawings might represent a form to "reach" them in some ritual form. Moralizing gods (high gods) are excluded from such construction, linking them as a reflection of the subsistence, social and political structures in whose bosom they originated, according to Peoples and Marlow (2012) and Peoples et al.

(2016), associated with, or preceded, by ritual practices affecting large populations (Whitehouse et al., 2019).

Symbolic thought would have emerged after *H. sapiens* speciation. In this sense, it is interesting to quote Tatersall (2010):

> "...was well established long before any population of it began to show indications of behaving symbolically. This places the origin of symbolic thought in the realms of exaptation, whereby new structures come into existence before being recruited to new uses, and of emergence, whereby entire new levels of complexity are achieved through new combinations of attributes unremarkable in themselves. Both these phenomena involve entirely conventional evolutionary processes; special as we human beings may consider ourselves, there was nothing special about the way we came into existence."

7.3. Rituals, Shamans, and Social Cohesion

> "Shamanism, ancestor worship and animism all entail an extension of the social world into a supernatural realm. This supernatural layer of social life brought with it a "spiritualizing" of social scrutiny – ever-vigilant eyes constantly monitoring for proper behavior, adherence to tradition, and avoidance of taboo" (Rossano, 2009).

From a comparative view, it is interesting the following remark by Rossano (2006):

> "There is no compelling evidence that non-human primates have religion. However, some foundational elements of religion such as ritual behavior and social emotions are present in our primate cousins."

As suggested by Peoples et al., (2016) (see also Dunbar, 2003; Coward, 2015), rituals and collectively mystic behavioral expressions would possibly have an ancestral origin:

> "The universality of religion across human societies (Brown 1991) suggests a deep evolutionary past. ...and some aspects of religion may have been present before the appearance of anatomically modern humans (2006) ... animism was the earliest and most basic trait of religion because it enables humans to think regarding supernatural beings or spirits. Animism is not a religion or philosophy, but a feature of human mentality, a byproduct of cognitive processes that enable social intelligence, among other capabilities... This innate cognitive trait allows us to attribute a vital force to animate and inanimate elements in the environment... Once that vital force is assumed, attribution of other human characteristics will follow... Animistic thinking would have been present in early hominins, certainly earlier than language."

It seems opportune to insert here another view on how *H. sapiens* might have coped with fear and uncertainty toward threatening events. It was mentioned before that in the animal kingdom, the possible responses to a threat is to fight, flight or freeze. To these, *H. sapiens* added animism, a subjective way to deal with fear, and to attempt to ease up the relationship with the unknown. In this case, another – virtual – dimension was added to the physical world that with time conquered their imagination in various forms.

It is interesting that several authors observed the existence of ritualistic behaviors in non-human primates (Brüne, 2006; Boyer and Lienard, 2006, 2008; Eilam et al., 2010; Eilam, 2015). Boyer and Lienard (2006, 2008) propose that such ritualistic behavior is the expression of a *Hazard Precautionary System* directed toward the detection and reaction to *supposed risks* – as different from *actual risks* that induce flight behavior.

In ancient ages, at primal times of the genre *Homo*, the relationship of the individual with the environment must have been severely affected by potent natural phenomena, which triggered emotional states linked with uncertainty, fear and a sense of fragility that would also have affected the group. The need to build up a constructive and manageable relationship with such forces and to glue intragroup relationships probably triggered cognitive and emotional individual and group adjustments and productive behaviors. Attribution of intentionality and agency to various objects and animals, as well as to "low" and "high" gods, must have eased up those feelings. Building virtual roads psychologically made them less distant or unreachable, and the possibility of "making deals" within their earthly, intimal realm. Interestingly, Botero et al., (2014) conclude that certain beliefs in "moralizing gods" tend to promote cooperation when more prominently poor or deprived is the environment and greater the risk of exposure to environmental stress. Given the strong correlation observed between cooperative behavior and ecological uncertainty, Botero et al., (2014) conclude that their observations support the notion of an existing link between ecology and religion (or beliefs in general):

> "Given the strong correlation between cooperation and ecological uncertainty in nonhuman animals (2, 3), these findings are especially suggestive of a link between ecology and religion."

It seems pertinent to include here some Piagetian observations concerning initial infantile stages related to the attribution of life or intentionality to inanimate objects (Piaget, 1928); an attribution that it could take origin on the lack of differentiation between the "I" and the "world". It could be said that the initial stages of *Homo* evolution would have transited through similar instances, and later persisted in such initial, animistic, perception of the world. This would indicate a persisting emotional immaturity of our species to which virtual or fantastic elements and physical representations were added, resulting in the construction of a religious "incarnation".

In this context, animism – the attribution of a vital force and intentionality to geo-climatic phenomena and inanimate objects – would have been the first rationalization of emotional constructions possibly related to the development of the *Theory of Mind*. Progressively, such vital attribution incorporated human features – a somewhat "humanization" of inanimate objects and natural events – later involving community members, whether alive or dead. Rossano (2006, 2009) poses a possible developmental chain starting with animism and progressing toward the emergence of complex religious constructions.

Several studies mention the ecological context in which the relations between *Homo* and its environment became a source of beliefs (see e.g., Peoples and Marlow, 2012; Xygalatas et al., 2013; Botero et al., 2014; Peoples et al., 2016), depending on the productive structure of the community after the stage of hunter-gatherers. Such studies of an anthropological nature point to the changes in ritualistic and other imaginary constructions, going through shamanism and witchcraft, until their eventual replacement by poly-or monotheism. That is, by religious constructions that, depending on social context, incorporated moralizing and punitive threatening gods, progressively imposing sophisticated rituals, temple buildings and a stiff hierarchical structure in the religious organization. This process ended with the institutionalization of beliefs, parallel to those belonging to the community structure, and depending on its productive and socio-political characteristics, particularly concerning almighty gods (high gods) (Peoples et al., 2016; Botero, 2014). In most cases, the objective was to generate a profitable relationship with an imaginary world occupied by forces or creatures – whether anthropomorphic or not – that supposedly control environmental events, as well as the fate of human beings. A complex religious culture progressively ignited the imagery of artists, resulting in productive creativeness involving sensible fine-art creations that emotionally appealed and recruited further believers.

In general, such constructions inspired a prosocial behavior based on love and fear, of group cohesion even implementing forms of sacrifice – animals and humans – as a proof of submission, promises of behavioral improvements or punishment for guilt.

7.4. Individual and Collective Ritual Behaviors

> "Anthropologists and sociologists have speculated that rituals involving synchronous activity may produce positive emotions that weaken the psychological boundaries between the self and the group...

endowing some cultural groups with an advantage in societal evolution" (Wiltermuth and Heath, 2009).

Complex, stereotyped and repetitive behaviors take place in the whole animal scale (Eilam, 2015; Eilam et al., 2010; Kuhl et al., 2016) in conditions of extreme alert state – generators of anxiety –, among them, during reproductive competence, predator risk, and leadership competence. They share a common trait that functions as a signal communication generator (Lorenz, 1966). This constitutes a common feature in the ritual expression of different species, including humans. These behaviors are considered ritualistic whenever they are repetitive and constant in its expression and structure or display. Nevertheless, Wightman (2015) establishes a difference between routine and ritualistic behaviors:

> "Though some may debate the point, it can hardly be said that chimpanzees – and by inference our last common ancestor – entertain beliefs that we can call religious. Though like all organisms, chimpanzees engage in some forms of behavior that are patterned and more or less routinized, they do not make the important transition from routine to ritual; that is, they do not symbolize their routines and elevate them to a general order of meaning. Nevertheless, chimpanzees and other Great Apes do engage in some forms of behaviors that also underpin human religiosity...However, they do not symbolize, generalize, and culturally transmit their individual experiences and behaviors."

Some additional comment seems justified to be included here regarding the use of symbolic terms, in the context of Wightman (2015). In the various animal species there exist different forms of expressive states of mind and intentionality. These exhibitions (gesture, vocalizations), although they may acquire a ritualistic form such as in situations of presumed threat, competition or reproductive courtship, they have the inherent value of coded information. Hence, they would include some form of symbolic expression, to the extent that they express the intentionality of achieving a behavioral objective. Such an objective could be proper or addressed to its community in case of a hierarchical attempt or of an impending threat. Exhibitions probably are also triggered in situations that activate the proposed *Hazard Precautionary System,* mentioned earlier. Behaviors described by Tinbergen (1965, cf. Fabricius, 1966) as *fixed behavioral patterns* are triggered at given sounds or aerial images (of predator threats) whenever they acquire a minimal threshold configuration, which represents coded information to induce alertness or flight. Although this type of behavior lacks the flexibility and adaptability of true language forms, they still operate as coded communication. A *Hazard Precautionary System* also seems to be operational in interpersonal relationships among humans, based on threatening gestures or past experiences based on threats.

In the case of humans, for Boyer and Lienard (2006), ritualized behaviors would be secondary products of a cognitive architecture and therefore would represent an indirect consequence of their evolution by natural selection. Such type of behavior, with gestural redundancy and repetitive, predictable movements would tend to confer a sense of control and prediction of anxiety-generating events for alleged potential threats (Eilam et al., 2010; Lang et al., 2015). Besides the complexity of certain human rituals, the underlying substrate mimics those expressed in other species, triggered by the *Hazard Precautionary System.*

These socially accepted collective behaviors oriented toward a clear objective must be differentiated in its phenomenology from the ritual behavior observed in cases of *Obsessive Compulsive Disorder* (OCD), although its construction could share some subcortical brain circuits.

This type of behavior, whether in normal or pathological (OCD) conditions recruits similar neural circuits that involve subcortical ganglia (striate body) (Graybiel, 2008; Martos et al., 2017). Neurocognitive models explain OCD pathology on the basis of hyperactivity of the *Hazard Precautionary System* (Szechtman and Woody, 2004; Woody et al. 2005; Boyer and Lienard, 2006), or *Evolutionary Threat Assessment Systems* (ETAS) (Flannelly et al., 2007), with involvement of basal ganglia, prefrontal cortex and limbic system, or at the extreme of a continuum of behavioral and cognitive strategies to avoid damage (see in Brüne, 2006). Several authors have demonstrated the participation of basal ganglia in behaviorally related displays in case of reproductive or territorial competitivity or in behavioral routines of grooming, courtship, and aggression. In humans, ritual behaviors can be observed when entering a challenging arena for competition, in religious circumstances, and in simple individual rituals to avoid some form of magical punishment due to a given mishap. In such cases, processing could be based on emotional contents, in which case there is a fast response; automatic and probably unconscious, and does not require language; or of cognitive nature, in which case language is required. Flannelly et al. (2007) summed it up that beliefs affect processing of data from reality and proposed that could be due to the involvement of prefrontal cortical circuits that receive afferences from various neuronal nets involved in the processing of other dimensions. For several authors, this comparative phenomenology of ritual behavior represents an evolutive basis of the emergence of religiosity.

In such respect, Rossano (2006) concludes:

> "...All of this suggests that the foundations of religion and religious behavior were laid deep in the primate brain possibly prior to the onset of hominid evolution."

Hence, according to these authors, it could be concluded that religiosity contains elements of a foundation deeply rooted in our animal evolution. These ritualistic behaviors are considered to be pathologic when they alter or interfere with normal behavior (Martos et al., 2017); otherwise, they would have evolved as an emotional, adaptive factor once the virtual dimension entered the notion of "reality", as a derived construct of cognitive and emotional development.

7.5. Rituals, Moralizing Gods and Religious Practices

Botero et al. (2014) analyzed these expressions in terms of the existing correlation between environmental uncertainty and cooperative forms, such as those induced by ritualistic behaviors. The progressive development of beliefs based on fantastic constructions, "moralizing agents", and their impact on community cohesion lead to the presence of "proxies" – shamans, initially. These became bridges between the earthly world and beings of the "ultra-world" – proxies that became social leaders with power over the basic emotions of their community and their "worldly behavior" (religious and political). In this respect, Peoples and Marlowe (2012) suggest:

> "...that belief in moral High Gods was fostered by emerging leaders in societies dependent on resources that were difficult to manage and defend without group cooperation. These leaders used the concept of a supernatural moral enforcer to manipulate others into cooperating, which resulted in greater productivity."

It should be noted that shamanic "proxies" with time were replaced by material icons representing some form of agency – talismans, religious cards with images, ritual objects such as prayer beads.

These effects of beliefs and rituals on social cohesion have been pointed out by several authors. Xygalatas et al. (2013) sustain the concept that *extreme rituals* amplify prosocial attitudes and behaviors to the point of generating a condition of generalized alert – as evaluated by changes in heartbeat frequency – among active participants and spectators (Konvalinka et al., 2011). Those events also reinforce group ideologies from which it is intended to exercise social dominance (Sidanius et al., 2004):

> "...the acceptance of ideologies that legitimize inequality and behaviors that produce inequality is partly determined by people's general desire for group-based dominance. This desire for group-based dominance is captured by a construct we call social dominance orientation."

On the other hand, Purzycki et al. (2017) highlight the impact of religious components on selfish behavior and control of sociability. These statements support the opinion of Peoples et al. (2016), in the sense that the profile of the content of the belief on moralizing gods would depend on the social and political construction of the community:

> "The results suggest that belief in an afterlife, shamanism, and ancestor worship evolve in concerted fashion as an integrated system of beliefs and practices. However, neither high gods nor active high gods exhibit correlated evolution with the rest of the religious traits, including ancestor worship... If a society acquires a belief in an omniscient and potentially morally punishing creator deity, it does so regardless of other aspects of its religion but more as a reflection of its social and political structure."

7.6. Possible Genetic Components in Religiosity

The neurocognitive field of religiosity is on a growing stage and supposes the consideration of several contents and expressive forms.

From an evolutionary point of view, based on its almost universal expression, it is interesting that several authors have suggested genetic participation in religiosity, though not of religious affiliation (D'Onofrio et al., 1999; Steger et al., 2011). Studies in mono- and dizygotic twins have provided a useful methodology for the analysis of the relative incidence of genetic and cultural components. These studies suggest the presence of hereditary components in the expression of religiosity, while religious affiliation would depend on cultural factors. Methodological aspects bring up debatable questions (see, e.g., Bouchard and McGue, 2003) and the need for large populations to validate them. The variety of components that characterize religious expression make many studies conceptually weak, as it is the case of the studies performed by Steger et al. (2011) on religiosity due to the limited number of cases. Studies based on mono- and dizygotic twins and adopted children suggest the presence of a hereditary component among the diverse components that characterize religiosity/religious attitude (beliefs in imaginary beings with intentionality, intensity of religious practices or rituals), being manifest or intimate (external- or internal religiousness) (see Bouchard et al., 1999; Waller et al., 1990; Likken et al., 1993; D'Onofrio et al., 1999; Bouchard and McGue, 2003; Koenig et al., 2005, 2008, 2009; Vance et al., 2010; Lewis and Bates, 2013; Vance, 2014).

According to D'Onofrio et al. (1999):

> "Results indicate that religious affiliation is primarily a culturally transmitted phenomenon, whereas religious attitudes and practices are moderately influenced by genetic factors."

Among other authors, this opinion is shared by Bouchard and McGue (2003):

> "Recent twin studies of adults, including twins, reared apart, and one adoption study, suggest that social attitudes such as Religiousness, Conservatism, and Authoritarianism are perhaps as heritable as personality traits. Religious affiliation, on the other hand, is shaped almost entirely by environmental influences. A more limited number of studies suggest that work attitudes and values, as well as job satisfaction, are also influenced by genetic factors."

Also, according to Vance (2014):

> "While it may at first appear that religiosity would not be influenced by genes, studies have shown that genetic effects contribute to individual differences in a wide array of traits and behaviors, including social attitudes, personality, vocational interests, IQ, and religiosity. The idea that religiosity is a heritable trait is not a new one. In his 1869 book Hereditary Genius, Sir Francis Galton wrote that "… a pious disposition is decidedly hereditary" (Galton 1869/1978, p. 274).

It seems clear that Galton's opinions were founded on evidence drawn from cultural influences, and not on genetic studies with methodological support.

Interestingly, most studies report that the expression of genetic factors related with religiosity tend to manifest during adult life, while cultural influences express themselves at younger ages (Koenig et al., 2005; see in Alexander, 2017). Would such a supposedly heritable trend be related to the *Hazard Precautionary System?*

7.7. Strategies and Neurocognitive Components in the Construction of Beliefs and Rituals

The theoretical "models" of the brain organization and those that try to formalize the concept of "religiosity" are a reason for continuous debate and contributions from the field of neurocognition. The following comments, as well as the previous ones, must be included within this dynamic knowledge.

As mentioned previously (Schiefenhövel, 2009), the construction of religious thought probably recognizes a distant origin associated with the neurobiological evolution of the cerebral cortex and mental processes, and the capacity to construct collective imaginaries. In this sense Inzlicht et al. (2012), like other authors (Barrett, 2011), speculate about the adaptive value of religious thought and rituals:

> "This model holds that people are motivated to create and sustain meaning (i.e., a sense of coherency between beliefs, goals, and perceptions of the environment, which provides individuals with the feeling that the world is an orderly place), and that religious beliefs buffer the distress associated with disruptions to meaning, thus leading to decreases in distress" (Inzlicht y cols., 2012).

This interpretation would provide an adaptive vision fundamentally in times of crisis, whether of collective or individual nature.

What neurocognitive processes give cohesive social power to religious rituals? According to Deeley (2010), religious rituals would employ two basic strategies in order to present them with a high sense of reality and an effective component. One, of "sensory" order, implements different practices with physical commitment. The second, of "semantic" order, uses verbal and non-verbal symbols intended to incorporate meaning into what is presented as real in authoritarian form, but which is not fully understood. Following the same author, both routes would activate, among others, components of the mesolimbic dopaminergic system in order to provoke states of mind identified as realities. According to Deeley (2010), these social and neurocognitive processes would be how religious ideas are transformed into conviction.

The participation of dopaminergic circuits (which include ventromedial components of prefrontal and temporal location) in religious behaviors has also been pointed out by Previc (2006). Several neural substrates of religious experience have been proposed; this would recruit areas associated with social cognition (Schjoedt et al., 2009). There are points of convergence in that the "sense of presence" of a supernatural being coincides with specific patterns of electrical activity in areas of the temporal lobe (Persinger, 2001), and it would also include prefrontal-parietal areas (Azari et al., 2001). Among other authors, Previc (2006) and Schwartenbeck et al. (2016) highlight the role of dopaminergic mechanisms in the updating of beliefs, associated with cognitive flexibility, and in aberrant behaviors (psychosis) with hallucinations.

This functional "proximity" between normality and mental states altered by shamanic induction and rituals, hallucinogenic drugs or schizothymic or psychotic states (Horrobin, 1998), represents a field for speculation about the origin and reinforcement of individual beliefs and cults (see also later).

On the other hand, Howlett and Paulus (2015) distinguish between empirical, verifiable beliefs, and non-empirical beliefs (ethical and religious), not verifiable. Its analysis using functional magnetic resonance revealed that while the first ones generate activation of the lateral prefrontal cortex and the posterior cingulate cortex, the second ones included the activation of areas within the inferior frontal gyrus and superior temporal gyrus. This differentiated recruitment of cortical neural circuits reflects the configurational cerebral counterpart of both types of beliefs. Their neurocognitive plasticity – modification by experience or learning – will be susceptible to be affected by different processes: the first one based on objective data, the second on experiences of virtual constructions. The dysfunction of these circuits and associated ones will generate different psychiatric pathologies with hallucinogenic manifestations and hyper-religiousness in case of temporal circuit lobe compromise.

In these models of cerebral functioning, some conceptual convergences arise on the construction of mental states as those that probably underline religiosity. Fundamentally, according to Barrett (2000), the construction of religiosity would be the product of ordinary cognitive processes and, therefore, would not depend on a specific nervous "center" but on a network of circuits involved with the "social brain" (amygdala, prefrontal orbital cortex and temporal cortex, according to Brothers, 1990) and the *Theory of Mind*. Additionally, it would include the over-detection of agents with intentionality in the environment related to circuits that make up the *Hazard Precautionary System* (Brothers, 1990; Schjoedt et al., 2009; Sztajer, 2012; Riekki et al., 2013; Neubauer, 2013; van Elk et al., 2016). Such circuits would involve neural networks present in the medial frontal and dorsolateral prefrontal cortices, the anterior cortex of the cingulum, the ventromedial temporal cortex, and the insula (Azari et al., 2001; Previc, 2006; Seitz et al., 2009; Schjoedt et al., 2009; Harris et al., 2009).

As can be seen, both the neurocognitive and comparative anthropological approaches have contributed to enriching the debate and the analysis of the probable origin and dynamics of these mental processes.

Chapter 8

FURTHER COMPARATIVE ANALYSIS OF CULTURE GENERATION AND THE EVOLUTIONARY ORIGIN OF BELIEFS (II)

Considerations on the Emotional Adolescence of the Species on the Possible Origin of Beliefs and Rituals, on Environmental, and Neurobiological Components - Rituals and the Hazard Precautionary System - On Religiosity[25] *- Social Dominance - Dominance and Emotional Dependence*

Social construction results from the interaction of emotional and cognitive domains affecting individuals and their interactions. Among those, religiosity is not a minor issue since it is linked with emotional dimensions and has an impact on social cohesion.

In the next paragraphs, contributions related to an evolutionary vision of the probable origin of beliefs and rituals will be further considered. Its impact on social cohesion represents an effect accepted by practically all

---

[25] Belief in supernatural agents is possibly the most accepted concept of religiosity, which has been considered instrumental in social bonding (Durkheim, 1969; see Alcorta and Sosis, 2005).

authors, and it is considered instrumental to the support ("bonding") of the communities. However, for this author, the recourse to resorting to virtual constructions of supernatural beings with intentionality would represent an indication of the emotional adolescence of our species.

8.1. Are We Transiting Through the Emotional Adolescence of the Species?

The concept of "reality" is a mental construction in which perception is affected by the interaction with knowledge, creativity, and mnemonic processes. The *degrees of freedom* to construct it will be affected by the range of options involved. Supernatural beliefs generated from altered states of consciousness, rituals, and acts of faith (see Colombo, 2013) provide imaginary components that tend to bias such a construction toward subjective domains. Beliefs based on emotional experiences bind us cognitively and emotionally to imaginary events. These are based on constructions of the supernatural dimension, which tend to control and "tame" our "animal roots" and condition our ancestral fears of the unknown, but also limit our *degrees of freedom* as long as it does not include the dialectic dynamics of proof and counterproof. From a collective standing, its effect on "community adhesion" tends to reduce individualist tendencies that could affect social cohesion, and therefore feed into tendencies of intragroup cohesion and cooperative behaviors. An alternative would imply that these objectives could be achieved by methods that do not cancel rational processes and rather promote the intellectual creativity of human thought, instead of besieging it with fears (Colombo, 2017). *Social responsibility based on fear is fear, not social responsibility*. Under such conditions, creativeness is guided toward fantastic, supernatural imagery. As it is mentioned in a T. Hobbes quote by C. Sagan (1995): "*The fear of invisible things is the natural seed of what each one calls for religion*". The question then remains; is this "supernatural" resource linked to an emotional, social immaturity of the species that fails to generate social models based on rationality without the need to resort to “supernatural vigilantes”?

8.2. Environmental and Neurobiological Events in the Possible Origin of Beliefs and Rituals

The emotional dimension permeates all the acts of our executive and introspective behaviors. In the modern world, its role even in economic-financial decisions has been the subject of research in the field of neuroeconomics. The degrees to which emotional components affect decisions of the “executive” module justifies incorporating into this essay several considerations regarding the ancestral origin of animistic constructions and beliefs, due to their emotional content and their impact on the structure of gregarious communities organized hierarchically. The fact that beliefs/religions act as a buffer against the different stressful causes of life (see also Atran and Norenzayan, 2004; Inzlicht et al., 2012) – events that tend to generate the need to believe –, as well as its powerful social adhesive effect, provide arguments for their cultural insertion and spread.

Nevertheless, when the emotional content of beliefs and ideas reach levels that cancel cognitive processing, fundamentalisms arise that generate a “shielding” to their questioning, an undesirable effect on dialectical capacity, an act of “social or group dominance or cultural predation”.

In this emotional dimension, it is worth asking why supernatural beliefs and rituals have acquired a generalized distribution and have been associated with intragroup cooperative instances, as well as with aggressive behaviors. These comments are not alien to evolved psychological characteristics of our species, such as the *Theory of the Mind.* This capacity allows us to assign interpretations about the state of mind of others and, by extension, can expand it to assign meanings and intentionality to natural events under an imaginary prism, constructions that would depend on a *Risk Precautionary System* (see above). To this ability must be added in the initial stages of the evolution of *Homo*, the emotional impact of altered mental states, dreams, and powerful geological and climatic circumstances. These would have inspired animistic conceptions and superior moralizing deities, which helped to build relations of subordination and dominance. On these emotional grounds were elaborated initial belief systems, rituals, and codes of individual and social ethics, which affected group cohesion and its economic impact such as productivity. According to Schiefenhövel (2009):

> "The early human mind, this is the position taken in this contribution, needed a powerful paradigm which was able to explain all the seemingly unconnected and unpredictable events of nature as well as its awe-inspiring beauty and functionality, and to make sense of the complexity of human wants and actions."

On the other hand, Harris et al. (2009) state:

> "According to Boyer (2001, 2003), religious beliefs and concepts must arise from mental categories and cognitive propensities that predate religion – and these underlying structures might determine the stereotypical form that religious beliefs and practices take. These categories relate to things like intentional agents, animacy, social exchange, moral intuitions, natural hazards, and ways of understanding human misfortune."

One way to deal with the unknown was represented by the shaman, a healing leader and intermediary of "supervisor and penalizer" beings who would inhabit supra- and infra-worlds. These relationships crystallized later into complex institutional structures that acted as proxies for such moralizing and punitive deities and affected the dynamics of relationships within and between communities. In short, religious beliefs underwent a process of earthly institutionalization, becoming instruments of social cohesion and regulation/control, of social policies and, in various circumstances – in their fundamentalist expression – mobilizing aggressive, repressive and belligerent behaviors, in short, they became territorial, returning to primitive drives of our animal origin.

8.3. AN EVOLUTIONARY COMPARATIVE EMERGENCE OF RITUALS

As can be suspected from the initial paragraphs of this chapter, the anthropological and neurocognitive analysis of the emergence of religions is a field of debate among specialists. However, what mental conditions and brain substrates could serve as a bridge to a collective influence, based on a supposed immaterial, virtual reality?

According to Tomasello (1999):

> "The basic puzzle is this. The six million years that separate human beings from other great apes are a very short time evolutionarily, with modem humans and chimpanzees sharing something on the order of 99 percent of their genetic material. There is only one possible solution to this puzzle. That is, there is only one known biological mechanism that could bring about these kinds of changes in behavior and cognition in so short a time, whether that time is thought of as 6 million, 2 million, or one-quarter of a million years. This biological mechanism is social or cultural transmission, which works on time scales many orders of magnitude faster than those of organic evolution. One reasonable hypothesis, then, is that the amazing suite of cognitive skills and products displayed by modem humans is the result of a sort of species-unique mode or modes of cultural transmission."

Interestingly, the complex cerebral substrate necessary for the emergence of imaginary, fantastic creations with an impact on individual and collective behavior was already expressed in non-human primates (see studies with images in Neubauer, 2014; Riekki et al., 2014).

Alcorta and Sosis (2005) draw a comparative bridge with ritual behaviors in non-human animals:

> "Like the elicit neurophysiological responses in participants and influence the nature of the social interaction (Lewis and Gower, 1980; Reichert, 2000; Rogers and Kaplan, 2000; Sapolsky, 1999)."

These authors add later:

> "EEG recordings have shown that the driving effects of ritual, such as music, drumming, and dancing, are capable of altering neuronal firing patterns in nonclinical populations, as well (Lex 1979; Neher 1962; Walter and Walter 1949). Elements of religious ritual that increase neuronal firing rates prime ritual participants for the conditioned association of symbols and emotions, both positive and negative, and create communal conditions for investing religious stimuli with these emotions (DiChiara 1995)."

Studies performed in non-human primates and other species indicate sharing of several primary neurobiological components in the generation of beliefs, although far from the richness and complexity with which they are expressed in humans. These are involved in some variable measure in processes and behaviors such as those in the *Theory of the Mind*, ritualistic behaviors, in cases of conflict conditions, and following consumption of plants with effect on the state of consciousness and altered behavior (see Rossano, 2016; Hare et al., 2000; Samorini, 2002 – cf. Rossano, 2016). These observations support the notion that there is a bridge or primary evolutionary, neurocognitive and emotional antecedent, although our distant ancestors must have expressed these components even more richly – complexity – and purposedly than non-human primates. The latter would have been expressed mainly in the second migration group from Africa of the *anatomically modern human* (see below).

According to Rossano (2009), the presence of the *anatomically modern human* (AMH) in the *Levante* region would have occurred in two phases, separated by a cataclysmic event (eruption of the Toba volcano, a relationship questioned by some authors), which would have decimated the AMH population of the the first migration. The second migration would have occurred 60,000 years ago. It would have been carried out by tribes of similar individuals but, unlike the first migration that occurred 100,000 years ago and extinguished 30,000 years later, collectively different, linked by strong cooperative ties and with a more significant social complexity of beliefs. The latter would have allowed a more successful encounter with the Neanderthals who inhabited the region.

Beliefs are strongly related to the over-detection in the environment of the presence of agents, i.e., vital forces with intentionality (e.g., random movements or sounds in the natural environment that trigger alertness in several species) (van Elk et al., 2014). Interestingly, this interpretation is associable to the observation that believers – rather than non-believers – tend to assign intentionality to randomized movements of geometrical forms (Riekki et al., 2014).

Although there is no accessible evidence that non-human primates (chimpanzees, bonobos) have beliefs, they do express a series of ritualistic behaviors and expression of social emotions. These are aimed at generating trust and intragroup cohesion or expressing welcome to previously split groups, which constitute elements foundational for religious thought, as described by several authors (see Rossano, 2006, Atran and Henrich, 2010). According to Alcorta and Sosis (2005), the differentiation of human rituals with those expressed in other species resides in the emergence of symbolic constructions with a strong emotional charge.

In conditions of collective risk, the mechanisms of social cohesion and coalition formation represent a productive response evolutionarily conserved in the gregarious communities. Such relations are reinforced with ritualistic behaviors and also destined to summon vital forces expressed in different fantastic constructions; whether virtual or material, in this case, represented by artifacts. According to Atran and Heinrich (2010), Schloss and Murray (2011), Xygalatas et al. (2013), Peoples and Marlow (2012), the rituals – particularly the intense rituals that reach physical risk – and religions have cooperative and social cohesion effects, as well as generating enmity between groups (Atran and Heinrich, 2010). The intensity of this "religious parochialism" can generate belligerent fundamentalisms and confrontational events as a secondary form, under the supposed consent of a protective god as the past and recent history of our civilization demonstrates.

The issue of "culture" in a comparative approach

According to Whiten et al. (1999*): "The extensive, multiple variations now documented for chimpanzees are thus without parallel. Moreover, the combined repertoire of these behavior patterns in each chimpanzee community is itself highly distinctive, a phenomenon characteristic of human cultures but previously unrecognized in non-human species."* Horner and de Waal (2009), support the idea that non-human animals can share similar cultural, formal expressions as proposed by Kinji Imanishi (1952). According to those authors, cultural differences between groups should not respond to genetic or ecological (habitat) reasons. For Whiten (2017), it constitutes a second hereditary system which interactively conditions genetic expression. The comparative evolutionary approach of these topics allows developing inferences about the possible collective and individual constructions in the *ancestral Homo*. However, its definition generates diverse opinions about its degree of inclusion. Although some anthropologists and psychologists deny the existence of culture in tribes of chimpanzees by assigning characteristics that would only apply to human culture, Boesch and Tomasello (1998) consider that there are cultural processes – traditions – shared with other primates. According to these authors, the comparison of human and chimpanzee cultures shows a series of profound similarities, suggesting that they share evolutionary roots. Two main differences are pointed out by these authors. These are the language impact in its diffusion, and the accumulative characteristic that allows the generation of elaborate cultural artifacts – instruments – in the *Homo sapiens* case. This would also explain the different evolutive impact of culture between these species.

Henrich and Tennie (2015) define "culture" as the emerging product of several generations and ages of individuals that learn interactively. Moreover, they add that there are several lines of evidence that support the idea that chimpanzees express "social learning", although with a reduced fidelity of transmission and without expressing a path of genetic and cultural interaction like humans. As a further differentiating developmental event, the prolonged period of neoteny in humans must be added which allows early access to social interaction while the brain is still developing.

The inclusive nature of the concept of "culture" proposed by some authors restricts its application only to humans. For example, Sahlins (1976), (see in Galeg, 1992), who maintains that culture would have developed in hominids about three million years ago. Whiten and van Schaik (2007) argue that "culture" represents the possession of several traditions (types of behavior between two or more individuals in a social context which lasts over time). This concept encompasses several dimensions of behavior; its presence is still under discussion in species in the wild but tends to be accepted in the case of the great apes and Capuchin monkeys.

According to these authors, there are correlations between social complexity, social learning capacity and the degree of encephalization which suggest the existence of an evolutionary spiral based on social learning and the generation of traditions. Therefore, an exclusionary view of the concept of "culture" adapted to humans would not allow comparative evolutionary analysis. In the concept of Galef (1992), animal traditions rely on processes different from those that support culture in humans, representing rather a homologous condition (similar structurally or formally), rather than analogous (functionally similar). For Ellen and Fischer (2013), the repeated discovery of the transmission of cultural materials by chimpanzees must call attention to avoid teleological definitions, although the greater the social and linguistic complexity, the more certain is Geertzian's (1973) definition of "culture" as a "network of meanings".

A couple of additional thoughts seem pertinent to close these commentaries: *"Cultural primatology is a breakthrough for understanding the complexity of our nearest living relations and for modeling the evolutionary origins of human culture." "... the remarkable human propensity for culture did not spring out of nowhere; research in decades has revealed much that it shares with the African apes and other species, providing unique insights into culture's evolutionary roots" (Whiten, 2017).*

8.4. Possible Evolutionary Origin of Beliefs: Impact on the *Degrees of Freedom*

Aspects of the possible ancestral conditions that gave rise to the generation and evolution of the different beliefs and religions are next analyzed. Based on the theoretical speculations and their possible link with evolutionary components of the productive profiles of gregarious communities, they will be considered regarding their impact on the *degrees of cognitive freedom* (Colombo, 2013) and their relationship with the social construction of the individual identity.

Approaching an answer to the issue of religious institutionalization would allow outlining the potential relationships between behaviors emotionally subordinated to beliefs and ideologies, and their transfer to institutional relations, as they may take place in the case of hierarchical structures of gregarious communities. Their dependence on emotional and cognitive dominance exerted by ritualistic and religious structures, to which a series of emotional states associated with fear are projected have conditioned other relationships.

In different periods of the development of our civilization, fear has been an instrument with which "taming" ancestral tendencies was also attempted, as these also affected creative proposals that threatened to fracture autocratic inertias and reformulate dynamic social stability. Judging from the collective development and with an optimistic spirit, this last one would seem to be part of an implicit Utopia. It is challenging to transition from "fear" to "freedom"; this generates a conflict of persistent characteristics unless one of them succumbs under the weight of the other.

The imprinting of "fear" to imagined worlds is not completely germane to traditional child rearing. This unfair practice complicates the "shaking off" of such a restrictive emotional corset, and to display optimized *degrees of freedom.*

The emotional dependence on ritualistic behaviors, magical constructions, or cognitive subordination in dogmatic constructions or of revealed truths, severely conditions the *degrees of freedom* required for a rational consideration of life events. The involved basic emotional configuration modulates the subordination or dependence on slogans/personalities of leaders, based on a fear of high gods or social contexts that affect the exercise of the citizen's executive behavior.

On institutionalized grounds, the emotional transfer – whether positive or negative – toward charismatic leaders who build an emotional relationship with the masses affects the cognitive demands essential for a rational consideration of this relationship. In these conditions, the intellectual distance with the assignment of an aura of empathy, which is constructed on a cluster of dissatisfactions, desires, expectations, beliefs, is reduced to terms that compromise the ability to discern. The dynamics outlined have representation in the concept of *group dominance* mentioned previously, or *cultural predation*.

8.5. Rituals and the Risk Precautionary System

Fiske and Haslam (1997) suggest that pathological behavior in the case of *Obsessive Compulsive Disorder* (OCD) represents an over-activity of mechanisms that generally operate in the case of social rituals (see also Brune, 2006). According to Brune (2006), the rituals associated with cultural manifestations are markedly similar to some components observed in the OCD, but the former is socially accepted as part of traditional rites. On the other hand, from the evolutionary and comparative point of view, Boyer and Lienard (2006) propose to differentiate between analogy and homology in the rituals expressed by humans and other animal species, as well as the existence of a *Risk Precautionary System*. This system would be aimed at detecting and reacting to supposed threats, a system that differs from that aimed at responding to manifest dangers. In this sense, the collective rituals would activate a cognitive-emotional system focused on the detection and reaction to *supposed* potential dangers. According to McCauley and Lawson (2002), what would be specific to a religious ritual is the incorporation of a supernatural agent(s). To please or to render accounts before said agents generates a condition of anxiety that is resolved after the execution of specific collective or individual ritual behaviors.

In the collective rituals, synchrony is essential as it confers powerful effects of group affiliation and increases prosocial behaviors which can connect thousands of people at the same time, as in the case of religious rituals (Wheatley et al., 2012). In particular, extreme rituals with different levels of suffering promote social cohesion and cooperative behaviors as well as empathy among observers (Durkheim, 1912; Konvalinka et al., 2011; Xygalatas et al., 2013; Fischer et al., 2013). The *Risk Precautionary System* would control the feeling of knowing the completion of the risk condition, which would generally cancel the thoughts or actions. In the case of OCD, there would be the inability to generate such a "sense of knowing" that would otherwise signal the completion of the task and end with the activation of the motivational safety system.

According to Rossano (2016), four conditions would be necessary for cognition construction of religious mentality. They are: 1. a system for detecting hyperactive agents – or organisms with defined purposes, such as protection – ; 2. a *Theory of Mind* – capacity to intuit other's mental state – ; 3. altered states of consciousness – trance states or mystics – ; and 4. imagination – capacity to create models of gods, spirits, souls, hells and other fantastic, supernatural concepts, not limited by the reality of the immediate present. In this regard, the analysis of Peoples et al. (2016), propose the evolutionary emergence of different religious beliefs based on observations in communities of hunters and gatherers: animism, belief in life after death, shamanism, adoration of the ancestors, and almighty gods. They affirm that:

> "...It can be inferred from the analyses, or indeed from the universality of animism that the presence of animistic belief predates the emergence of belief in an afterlife. Once an animistic thought is prevalent in society...The afterlife might be a rewarding continuation of life on earth, or a realm of eternal punishment for those who break social norms. Belief in an afterlife may have generated a sense of "being watched" by the spirits of the dead, prompting archaic forms of social norms (Bering 2006) actualized in the role of the shamán... Although shamanism has

been described as the universal religion of Paleolithic hunter-gatherer..., it is not a religion per se, but a complex of beliefs and behaviors that focus on communication with the ancestral spirits... Chamáns are healers, ritual leaders, and influential members of society..." (Peoples et al., 2016).

Several examples of animistic constructions that led to myths, legends, superstitions, and cultural traditions have been pointed out. It is interesting that urban culture incorporated several of them as "reinforcement" to other magical constructions. Among them are; throwing salt backward to ward off bad luck, the red ribbon against envy, the four-leaf clover to achieve good fortune, the use of amulets, belief in the *apotropaic*[26] effects of garlic strings, the ritual of crossing oneself several times before starting any competition, making strenuous pilgrimages and offerings to physical representations – a kind of idolatry – of religious icons. It has been posted in the press media the "*throwing of stones to the devil*" in a mid oriental ritual culture that gathered thousands of pilgrims (EFE, 2009; Garrido, 2018). This "marks" the emotional dependence on a magical dimension in the individual construction of the "real world" and displaces the need for ecological factors in the internalization of certain beliefs, as some authors maintain. Rather, it tends to point to the predominance of ancestral components of a psychological nature linked to a sense of insecurity/uncertainty and the search for "protection" or to provide a sense of reassurance regarding personal objectives. This susceptibility to the insertion of magical elements in urban and suburban cultures suggest that these constructions would be associated with ancestral tendencies in our mental organization affected by cultural factors and probably related to a *Risk Precautionary System*. They are evidence of the profound brain "*incarnation*" of rituals and magic motivations, supposedly with a protective effect.

---

[26] Anthropological term to describe a cultural phenomenon expressed as a magical or supernatural defense mechanism evidenced in individual acts, rituals, objects or formulaic phrases (https://es.wikipedia.org/wiki/Efecto_apotropaico).

8.6. Rituals as Beliefs Reinforcers. Social Dominance and Institutionalization of Beliefs

However, the central issue in our case is the possible relationship of such beliefs with the transfer of positions of dominance or dependence – subordination – in a community with a hierarchical structure. This may be in the projection of imaginary qualities that are possibly remnants of ancestral magical traditions built on a common substrate of insecurity/fear, or virtual constructions (gods) sometimes represented in geographical accidents or climatic phenomena or embodied in the worship of anthropomorphic representations. Alternatively, these may also be induced by hallucinogenic exaltation or oneiric of personal experiences. They sometimes take different forms of rituals in which self-punishment-flagellation, fasting, or strenuous pilgrimages are expressed as an offering and "proof" of such subordination. The objectives of these rituals may be directed to assure the personal well-being of the community or the predominance in the struggle of sects or fundamentalisms of various kinds. Transfer of special powers to mythical figures, often enlisted in epic constructions, delegates to them the exercise of guilt, merit, and punishment. The character of "proxies" is exercised by those who represent a supposed earthly connection through virtual constructions with those imaginary life forces endowed with supposed omnipotent powers[27].

It has been described that the nature of the rituals encourage cooperation and that the belief in high gods that punish morally contribute to prosocial conditions and social stability (Norenzayan and Shariff, 2008; Atran and Henrich, 2010; and Marlowe, 2012; Xygalatas et al., 2013); a "reinforcement" to the construction and imposition of values on individual behavior. Such beliefs generated cognitive and emotional conditions for subordination – delegation of *degrees of freedom* – to representatives of terrestrial (socio-political) or religious institutions with emotional dependence on the supposed power of amulets and rituals, who concluded

[27] Expressed at the level of the popular song by Bob Dylan in his work "*With God on our Side*".

administering conditions for the mental organization, establishing the admitted values and intimate freedoms of the members of said communities. It is interesting that Schjoedt et al. (2013) conclude that the cognitive effects of religious interactions tend to increase the susceptibility to authoritarian narratives and interpretations – ecumenic hierarchies –, preventing people from constructing their own experiences of the ritual event.

The theory of *social dominance*, which is a stage before that of *social predation*, focuses on individual and structural factors that contribute to the construction of various forms of oppression based on groups. Their different partial expressions are considered individual cases of a general tendency of humans to generate and maintain hierarchies based on groups (see Polimeni, 2002; Sidanius et al., 2004). According to these authors, this theory proposes to consider the continuous interaction of factors associated with nature and culture as an engine of human behavior. Also, the production and reproduction of social inequities, as opposed to considering it as a product of separate dimensions (psychological predisposition, social identity, social context, ideologies). At any rate, nature and culture are inclusive universes of those "separate dimensions".

With urban and suburban migrating processes, it is surprising how often myths from Andean cultures or the Mesopotamian NE have been incorporated into urban cultures in Argentina. Others have remained localized to the region of origin.

The transfer, or attribution of unique characteristics, especially in times of financial or social crisis or episodes of a health crisis, extends not only to imagined figures. They also include projects of social leaders who manage to construct a proactive image for groups of citizens and generate a collective state of emotional dependence, whether they are leaders of religious or laic congregations. In given cultural contexts, social leaders tend to emerge with an aura where imaginary powers melt with actual rational considerations. In these cases, collective emotional mechanisms tarnish rational identity, and the power – representation – of majority citizen groups is transferred or delegated to such leaders. This process usually generates movements with an extended dominant capacity that exceeds the rationale limited to the original community. In this sense, Sidanius et al. (2004) state that the *Theory of Social Dominance* focuses on both individual and structural factors that contribute to various forms of oppression based on groups:

> "Rather than being an exercise in psychological reductionism, social dominance theory is explicitly devoted to trying to understand how psychological predispositions, social identities, social context, social institutions, and cultural ideologies all intersect to produce and reproduce group-based social inequality" (Sidanius et al., 2004).

In this context, the emotional dominance posed by the major religions, citing the omnipotent power of moralistic gods, has a direct impact on the cohesion of the communities; an effect that in past times or in agrarian communities were tightly associated with their dependence on food production and physical defense, which required strong social cohesion. These goals were usually achieved by leaders who claimed to be "close" to supernatural forces. Such emotional dominance was transferred to the material dominance exerted by institutionalized structures. In this sense, the social and political organization of modern communities would influence the belief in omniscient creative deities, as expressed by Peoples et al. (2016):

> "The results suggest that belief in an afterlife, shamanism, and ancestor worship evolve in concerted fashion as an integrated system of beliefs and practices. However, neither high gods nor active high gods exhibit correlated evolution with the rest of the religious traits, including ancestor worship, despite Spencer's and Tylor's suggestions. This is in line with a variety of evidence from other studies (Botero et al. 2014; Norenzayan 2013; Peoples and Marlowe 2012; Radin 1937; Swanson 1960) suggesting that if a society acquires belief in an omniscient and potentially morally punishing creator deity, it does so regardless of other aspects of its religion but more as a reflection of its social and political structure."

Social dominance is reflected in the belligerent campaigns between groups or nations, where religious values are projected into militants and flags of opposing factions. From the anthropological point of view, it is interesting that such "taking sides" is carried out by emotional transfer to leaders who embody strong collective motivations (manifest destiny, supposed relative values, divine consent). Predatory activity on the emotional construction of people's identity is a phenomenon too close to that of emotional dominance. The annulment of emotional identity and its replacement by shared collective emotions are direct consequences and represent common strategies also applied in ideological and political arenas. In these instances, groups or crowds lose the critical identity of their components, diluted in a nebulous notion of belonging to an emotional collective.

In such situations, it is as if the capacity for cognitive discernment and the value of cooperative consensus in a dispute were suffocated by ancestral emotional inertias that bring us closer to parochialism and distance us from an integrating vision with social equity, closer to a human utopia.

To conclude, it is worth quoting several pertinent thoughts, such as that of Mauss (1902, translated in 1972):

> "... we suggest that a good part of all those non-positive mystical and poetical elements in our notions of force, causation, effect... could be traced back to the old habits of mind in which magic was born and which the human mind is slow to throw off" (Mauss, 1972).

Moreover, in the opinion of Alcorta and Sosis (2005):

> "... the critical element in the differentiation of religious from nonhuman ritual was the emergence of conditioned association of emotion and abstract symbols."

8.7. Beliefs and Social Dominance/ Cultural Predation

What we have described so far and the examples to be included below suggest some additional considerations based on two processes that may have developed progressively in bands of hunter-gatherers, tribes and institutionalized communities. One is the ability to transfer emotional states generated by fear, or fear to natural events of a significant magnitude or of unexpected nature, to fantastic cognitive constructions that ended up conditioning individual and community thoughts and behavior. Conditioning aimed at avoiding the supposed threat of uncontrollable forces to which were assigned power and terrestrial designs (by agents or moralizing gods or not). Leaders such as shamans assumed this animistic transfer who acted as proxies between the imaginary and the physical or real dimension. According to Eliade (1964), shamanism could be defined as a technique of ecstasy by achieving an altered state of consciousness through which one reaches a mystical state which provides a subjective approach to the deities of the supra- and infra-world. These agents reach an altered state of consciousness, whether through the intake of plants with psychotropic activity, the repeated sound of drums and rituals, hallucinations (acoustic or visual) and dreams, or delusions resulting from alterations of psychotic characteristics (schizophrenia, for example: see below). In these conditions, the "transference" of fantastic experiences, would become a condition for a collective "trance" that reinforces the belief in an immaterial dimension, with powers capable of controlling processes as distant and diverse as healing or geo-climatic phenomena. Those who dominated these "instruments", or built supposed "bridges" with the ultra-world, reached higher hierarchical ranks in the social organization. They progressively reached the status of intermediaries/interpreters with "supernatural" powers; gods that ruled over "good" and "evil" of communities or individuals. In the periphery of that range, amulets and idols were transformed into domestic proxies and icons of adoration and prayer – personal or collective – for the "communication" and the offering, in search of solving unsatisfied needs and aspirations.

The need to build conceptual structures that shield people from natural threats and death, generated gods, rituals, idols, amulets, compulsive behaviors, pilgrimages, and sacrifices of different sizes with the ability to pretend to rule or correct behavior or "destiny". From that imaginary world and emotional submission – later elaborated in a cognitive form – arose rules, ethical norms, commandments, institutions that regulated the admitted behavior. This mystical relationship with the unknown and fear-threatening agents provided a resource to modulate, regulate or tame societies that became progressively more complex and were pulled by tendencies inherited from ancestrally "embrained" behavioral trends.

In the inertial emotional development of our species, appealing to fears and expectations for the future is a weapon of extraordinary power; an effective resource of "conviction", potentially more potent than rational argumentation. Beliefs with an emotional basis are embodied in the brain and mental construction. A central emotional phenomenon in animal behavior has probably been critical in this process: insecurity, fear of real or supposed threats (see the adaptive impact of a *Threat Evaluator System* in Flannelly et al., 2007).

Faced with the impossibility of avoiding such threats, *H. sapiens* found a way to flee from or to reduce a conflict that surpassed its "emotional defense arsenal"; that is, the creation of a virtual dimension, configured and reconfigured to the extent of its cognitive capacity and inventiveness or imagination, and sustained by faith. This represented an act of surrender or submission of its rational autonomy in the construction of human development.

The emergence of beliefs and religions in different communities of the planet pose several dimensions for their analysis (philosophical, psychological or neurocognitive, anthropological or social). In this particular instance, the interest focuses on the possible evolutionary causes of cerebral/mental resources for its emergence. The association of the healing shaman and "interpreter" of gods of the supra-world – moralizing or not – would have crystallized in the form of complex forms of religiosity with social contents, and ethical values adapted to each social context that would satisfy alleged "demands imposed by moralizing gods". The "social permeability" to such demands would have acted as permissive conditions for its diffusion and mental "incarnation".

Convergent miraculous events in the construction of stories were reinforced by the regular practice of rites entangled in this narrative. Although counterintuitive, those constructions would generate emotional instruments for the containment and the rapid response to threats (*Precautionary System*) (Flannelly et al., 2007; Renfrew, 2008). These practices would find less resistance than those based on rationality, free of emotional content (empirical or verifiable beliefs according to Howlett and Paulus, 2015). The latter would require an "extra effort, an emotional countercurrent" (see discussion in Geertz and Markusson, 2010; Norenzayan and Gervais, 2013; Pennycook et al., 2016). As Barrett (2011) wrote:

> "For a theist, the attribution of agency is relatively simple: if detected agency cannot be attributed to any human or animal being, the agency may be attributed to a supernatural being. The situation is much more complicated in the case of an atheist who cannot interpret the detected agency in terms of superhuman agents. The only alternative for an atheist is to interpret the detection as a cognitive error."

If we take into account the previous considerations that show the cerebral "incarnation" of the animistic and religious expressions in the emotional dimension of the behavior organization, we should consider its impact on the *degrees of freedom of thought* (Colombo, 2013), even accepting the socially cohesive role of religions, or, as Atran and Henrich (2010) critically state it:

> "Both cross-cultural experiments and surveys indicate that people more readily ascribe to the veracity of narratives containing counterintuitive elements (e.g., miracles) when primed with death (Norenzayan and Hansen, 2006), or when facing danger or insecurity, as with pleas of hope for God's intervention during wartime (Argyle and

> Beit-Hallahmi, 2000). Such findings help explain both cross-national analyses showing that a country's religiosity (devotion to a world religion) is positively related to its degree of existential insecurity (Norris and Inglehart, 2004), and why certain kinds of religions enjoy revivals in challenging times."

Considering the rituals of non-human animals (see above) that suggest the expression of similar basic inertial behavioral tendencies – symbolically enriched in the human species – we should question, as an expression of our adherence to freedom of thought, the implicit cost concerning our *degrees of freedom*.

If we accept that from the emotional point of view, we would still be transiting an "adolescence stage" of human evolution, would the risk implicit in a weakening of the tendency toward "cohesion" as imposed by religious doctrines imply a greater social instability than the existing global one? However, we can also accept that the current and historical expressions of dominance, belligerence, fundamentalism, abuse, and corruption of political and religious power have ceased to represent an acceptable condition. Moreover, abandoning the "adolescence of the species" requires taking on greater individual and collective responsibilities. For this reason, perhaps it is the opportune moment to consider the forms of emotional dominance, or "*emotional autocracy*", should be progressively replaced by conceptions that turn to central, fundamental values of the human society; in a context of multicultural coexistence to the rationality of cooperation and solidarity. This implies the formulation and acceptance of new values and objectives as lucidly and dramatically were expressed by the Secretary of the United Nations, U Thant (1969), and the *Executive Committee of the Club of Rome* (1972) (see initial page). As stated earlier, "responsibility" based on fear is fear, not responsibility.

Chapter 9

SUPERNATURAL CONSTRUCTIONS AND ALTERED STATES OF CONSCIENCE (III)

We have reviewed the results of research and opinions about genetic, neurocognitive and cultural aspects regarding possible origins of beliefs and religiosity. However, paleoanthropological and biomedical investigations offer additional contributions toward understanding the probable origin of individual beliefs based on an imaginary realm; a dimension inhabited by beings with supposed omnipotent abilities, capable of regulating the events of the material dimension of the "earthly" events, and the personal circumstances of each terrestrial inhabitant. Let us briefly see some opinions.

> "A first definition of this complex phenomenon, and perhaps the least hazardous will be shamanism = technique of ecstasy" (Eliade, 1964).

> "It may seem questionable; nevertheless, magic is a technique in the strictest sense of the word, as has been demonstrated by Marcel Mauss. Magic developed along with other techniques as an expression of man's will to obtain certain results of a spiritual order" (Ellul, 1964).

"Religious concepts activate various functionally distinct mental systems, also present in non-religious contexts, and 'tweak' the usual inferences of these systems. They deal with detection and representation of animacy and agency, social exchange, moral intuitions, a precaution against natural hazards and understanding of misfortune. Each of these activates distinct neural resources or families of networks. What makes notions of supernatural agency intuitively plausible? This article reviews evidence suggesting that it is the joint, coordinated activation of these diverse systems, a supposition that opens up the prospect of cognitive neuroscience of religious beliefs" (Boyer, 2003).

"Because religious beliefs cannot be deductively or inductively validated, validation occurs only by ritually addressing the very emotions motivating religion. Cross-cultural experimental evidence encourages these claims" (Atran and Norenzayan, 2004).

9.1. Beliefs and Neurocognition

According to Kapogiannis et al. (2009), religious beliefs recruit neural networks involved in abstract semantic processing, imagery, and emotional aspects of *Theory of Mind.* These would be integrated into cognitive processes and neural networks involved in social cognition whose construction can vary in different cultures. According to Lewis-Williams (2016), the universality of beliefs in a supernatural dimension suggests that they would not depend on the intention to build an understanding of life, but in some way would be linked to brain organization. E.B. Tylor (1832-1917) proposed that the existence of the soul or a spirit – in short, of an immaterial existence – would have been suggested to our species' ancestors from altered states of consciousness (trances, visions or induced hallucinations or pathological, dreams). In the words of Lewis-Williams (2016):

"But the "wiring" is more comprehensive that he (Tylor) knew. As we shall see, features of dreams and some altered states of consciousness that are neurologically wired occur in most religions. These common

> features suggest that religion is closely related to mental states generated by the electrochemical functioning of the human brain... In summary, we can say that both neuropsychology and world ethnography show that the near universality of belief in a tiered cosmos and in movement between the levels (of consciousness) may be ascribed to the functioning of the human nervous system in a variety of altered states."

This concept of the possible origin of magical and religious beliefs from altered states of consciousness leads us to consider proposals that represent variations of such a concept; some of them with evolutionary projections as will be discussed later. Regardless of such a possible origin, when beliefs were institutionalized, they generated public policy instruments and established rules and values that affected the collective behavior of extended world communities. According to Durkheim (1972) (cf Lewis-Williams, 2016), religion (the belief in supernatural beings) would have an adaptive impact by providing means of personal and cooperative survival in hostile times. Concerning brain mechanisms as suggested by Lewis-Williams (2016) and the possible role of altered states of consciousness, these refer to dreams, hallucinations, and trances generated by the intake of natural products with psychotropic activity, employing music or repetitive and monotonic beats, or by mental pathologies (auditory hallucinations or visuals).

According to Burns (2007):

> "...a period which spanned perhaps 100,000 years, beginning with the very first fully modern *Homo sapiens*. Archaeological evidence from this period suggests that these early humans regularly experienced psychotic-like phenomena in the context of chamánic ritual practices and psychoactive substance use."

9.2. Beliefs and Altered States of Conscience

According to the same author (Burns, 2007) in the period mentioned in the previous paragraph, there are abundant records with the use of psychoactive plants for the induction of altered states of consciousness.

These conditions relate to the presence of shamans in communities of hunter-gatherers, a phenomenon that would have had widespread characteristics during the *Paleolithic* era.

The earliest evidence about the medicinal use of plants comes from a Neanderthal burial in Iraq dated between 50,000-80,000 years with large amounts of pollen from different plants, seven of which had medicinal properties (Leroi-Gourhan, 1975). It is interesting the comment of Burns (2007) regarding parietal paintings dated 12,000-15,000 years old found in Algeria. They decorated shamans carrying quantities of fungi, as well as graphics representing entoptic forms – visual effects such as flying flies, phosphenes – suggestive of altered mental states (McKenna, 1992). In Mexico, remains of human instruments from 8,000 years ago were found with mixtures of peyote and mescal. Burns concludes that the human of the *Paleolithic* era was capable of expressing hallucinations, mood alterations, and mental dissociation phenomena. For McKenna (1992), substances with psychotropic activity were part of the human diet during the *Paleolithic* period.

However, the incidence of altered states of consciousness in the genesis and expression of beliefs outside the material and the causal world is not limited to episodes of hallucinations, trances, and dreams generated by plants with psychotropic activity. To those factors should be added altered states of conscience due to a psychiatric pathology with productive behavioral components, whose expression adopts common characteristics in different regions of the planet and is of genetic origin. Both circumstances make it a candidate to be incorporated into an evolutionary social construction of our species, as proposed by different authors. It is about schizophrenia and schizothymic states. Its analysis has raised issues of comparative evolution, as well as debate in the field of neurocognition and the origin of beliefs, so much so that Kaufman (2016) wonders if hallucinations have had and still have a social impact. The author performed a historical analysis of personalities that expressed hallucinatory components and generated or were incorporated into religious beliefs and institutions or stood out in other aspects. Let us consider articles about it and its possible link with the theme of the generation of mystical/religious beliefs.

Although for some authors the hallucinations of schizoid pathologies could have generated ceremonies, others doubt it given the characteristical loss of social competence in this disease. Dein and Littlewood (2011) proposed the hypothesis that schizophrenia and religious construction share cognitive modules in assigning false attributions to ultrahuman characters and an over-dimensioning of the *Theory of Mind* and that they share an evolutionary trajectory.

Burns (2007) raises similar considerations regarding the hypothetical incidence of psychotic manifestations in the ancestral species common to *H. sapiens* and chimpanzees, similar to those expressed by the former, although depending on the cognitive condition and social ecology of the species. In his words:

> "Furthermore, there is strong data from comparative psychology, molecular biology and physical anthropology confirming the close evolutionary relationship between simian and ape species and modern *Homo sapiens*. So, since mental phenomena do not fossilize, this question is best addressed by examining the literature on our nearest extant relatives – the great apes."

Burns (2007) makes particular emphasis on repeated motor behavior ("rocking"), coprophagia, scatolia (spotting with feces) and pulling the hairs; behaviors described by several authors in monkeys and great apes. While there may be social learning, in many cases there are psychopathological manifestations that reveal a potential for psychotic expression, although associated with conditions of captivity. Based on the latter, it would appear that such pathological behavioral manifestations existed in prehuman *Homo* communities; their expression would have depended on – or triggered by – social dynamics.

Altered behaviors observed in great apes and monkeys often triggered by conditions of artificial habitat suggest in them the possibility of developing pathologies of psychotic nature, a consideration potentially extensive to our common ancestor.

The question, initially proposed by Premack and Woodruff (1978) (*"An individual has a theory of mind if I impute mental states to themselves and others")* of whether chimpanzees had the ability to capture the mental state of third parties' beliefs and desires, propelled the concept of the so-called *Theory of Mind*. From an evolutionary point of view, it has been proposed that the ability to interpret the mental state of the interlocutor would have originated between 5 and 16 million years ago (Mithen, 1996). This corresponds to a period of divergence of the chimpanzee and the Australopithecine/Homo lineage from a common ancestor, until achieving a marked development of the *Theory of Mind* in humans 40,000-150,000 years ago. There is a tendency to establish a difference with humans in the sense that, while they can elaborate a sophisticated *Theory of Mind*, chimpanzees would be limited to perceiving what other members perceive of the environment, but they would lack the ability to discriminate false beliefs. This would argue against an evolutionary process of the *Theory of Mind*, as suggested by (Baron-Cohen, 2012).

9.3. Evolutionary and Comparative Development of Cerebral and Mental Events and Their Relationship with the Emergence of Beliefs. Schizophrenia as One Possible Source of Altered States of Conscience

The studies on the evolutionary and comparative development of cerebral and mental events raise a series of questions. Particular emphasis has been given to the analysis of the ratio between brain volume and cognitive development on comparative grounds. In this sense, it is attributed to Holloway (1975) (cf. Burns, 2007) the argument that evolutionary changes in cognition reflect the reorganization of brain systems – essentially connectivity – rather than an increase in brain size, as several have proposed. Burns (2007) emphasizes the observations of

Hofman (1989) (cf. Burns, 2007) in the sense that white matter increased significantly relative to brain size, rather than gray matter. Moreover, intrahemispheric connectivity would have increased, rather than interhemispheric, which would have decreased with the increase in brain size (Rilling and Insel, 1999).

This increase in intrahemispheric connectivity would have occurred in particular between the prefrontal cortex and the parietal and temporal cortices, involved – together with the limbic system – in the so-called social brain circuits (Burns, 2004, 2007). This increase would respond to the pressure of the social complexity in the genre *Homo*, a progressive phenomenon according to Burns (2007) and not episodic as Crow (2008) maintains (see below). This concept of the role of social complexity, applied to its impact on brain size, has been proposed by Dunbar and Shultz (2007) in the so-called *social brain hypothesis*. This idea has been extended by other authors to incorporate the ecological and cultural dimension (Reader et al., 2011; Whiten and van de Waal, 2016).

Recall that, in the context of this chapter, the importance of considering psychotic pathologies such as schizophrenia resided on its inclusion related to possible sources of hallucinations and altered mental states. These could have been part of the generation of beliefs about a virtual world parallel to the material (see earlier in Kauffman, 2016) and expressions of hyper-religiousness. The paradox of the global survival of a pathological condition, such as schizophrenia which reduces fertility, has primarily been the subject of two interpretations.

According to Burns (2007):

> "The psychotic genotype itself was unlikely to have conferred a reproductive advantage on the schizotypy, but because of its association with genes that code for the development of the social brain in the species, the disorder persisted in the human genome. Thus, as Thomas Midwinter (Faulks 2005) proclaims 'The same "genes" that drive us may have made us human'."

Another of the scholars on the evolutionary aspects of this disease is Tim Crow (1997, 2000, 2008, 2013). According to this author, the evolutionary process that led to asymmetry in cortical representations and the development of language would be linked to a chromosomal transposition – mutation located in homologous regions of the sex chromosomes – associated with the emergence of language, speciation of *Homo* and its separation of the great apes (Crow, 1997, 2000, 2008). This process would be altered in the genotype of schizophrenia, resulting in disability to recognize the difference between their thoughts and the expression of their voice:

> "...it can be seen that the phenomena of schizophrenia and the problem of the origin of *Homo sapiens* are two sides of the same coin... The general conclusion is that the major syndromes of psychoses are constant features of human populations independent of culture and environment" (Crow, 2008).

On the other hand, Burns (2004, 2006, 2007) emphasizes the functional deficit of the nervous circuits involved with the "social brain" and incorporates the notion of "selective social pressures":

> "...the spectrum of psychoses should be regarded as a costly by-product of social brain evolution in *Homo sapiens*. Under selective social pressures, hominid ancestors evolved a sophisticated neural network supporting social cognition and adaptive interpersonal behavior – this is termed the 'social brain'. The functional psychoses (and schizophrenia in particular) express functional and structural deficits in these frontotemporal and frontoparietal circuits; hence the epithet 'social brain disorders' is fitting."

See also Berlim et al. (2003) who states:

> "Language originated by a genetic mutation that allowed the cerebral hemispheres to develop with a degree of specialization (or lateralization) reflected in cerebral asymmetries. Individuals with schizophrenia show lesser structural and functional brain asymmetries than the population as a whole, and this finding can be interpreted as a delay, or failure in, establishing hemispheric dominance for language."

The two hypotheses briefly described emphasize a shared observation referring to the universality of the nosological characteristics of schizophrenia that suggests a component linked to the evolution of the species. However, they differ in the possible reason why this pathology does not result in its elimination for reasons of maladaptation, as the Darwinian theory would suggest. For T. Crow, this would be due to genetic alteration in the sex chromosomes linked to cerebral asymmetry related in turn to the emergence of language and the process of speciation of *H. sapiens*. For J. Burns, schizophrenia is linked to genes shared by the "social brain":

> "In short, I argue that under the selective pressure of social group living, the primate brain first enlarged and then reorganized during the period 16 – 2 million years ago (mya). Complex neural circuits linking the pre-frontal cortex to the temporal and parietal cortices evolved as a substrate for complex social cognition. I also suggest that the ontogenetic mechanism underlying this cerebral adaptation... rendered the hominid brain vulnerable to genetic and environmental insults...However, because the disorder shared a common genetic basis with the evolving circuitry of the social brain, it persisted."

However, in addition to this interesting controversy about the neurocognitive evolution of schizophrenia (and associated syndromes), both arguments may not be mutually exclusive. It can be concluded that the productive manifestations of this dementia, expressed as altered states of consciousness, have been associated with the origins of our species. According to Dein and Littlewood (2011), religious cognition and schizophrenia recruit related mental modules, linked to the oversizing of acting entities and the *Theory of Mind*; a concept advanced by Polimeni and Reiss (2002), who point out similarities between shamanism, religion, and psychosis (also Thalbourne and Delin, 1994). According to the authors, the psychotic condition could have been advantageous for shamanic leaders by incorporating novel and persistent rituals from their altered state of consciousness. Therefore, hallucinations and other manifestations must have been socially shared events from ancestral times.

To this should be added delusions in febrile states, shared experiences of dreams or by consumption of products with psychoactive effects, as well as shocking geo-climatic phenomena that must have resulted in an active combination for the mental organization during birth and "early adolescence" of our species.

On the other hand, the different ritual practices, some still in force would act by breaking the boundaries between the "I" and the "other", or the surrounding world (Dein and Littlewood, 2011). This process fueled the emotional and cognitive construction of an immaterial, parallel world inhabited by omnipresent and omnipotent beings with an impact on destiny and the events of the material world. Some of these constructions became beliefs, myths and religious events (still valid in what the author considers the "late adolescence" of our species):

> "Hallucinations generated by psychosis were useful for cultural innovation, particularly in religion as many hallucinators were integrated into church history" (Kauffman, 2016).

These considerations indicate a condition receptive to the construction of virtual worlds parallel to the material world, inhabited by agents with special powers during the initial stages of the species. With their cognitive development came a second phase of the rationalization of these beliefs. They induced the recruitment of new, cognitive cerebral circuits along with the emotional ones, including prefrontal areas and other distributed circuits, according to the neurocognitive analysis about religious beliefs (see for example in Boyer, 2003; Dein and Littlewood 2011; Willard and Norenzayan, 2013; Kapogiannis et al., 2014).

Chapter 10

FINAL COMMENTS

The evolutionary and comparative analysis relative to the emergence and characteristics of our species reveals its primary biological nature. Its genomic pedigree combined along a chain of ancestors and, since its primeval construction, shared commensalism with prokaryotic organisms of a different order, depending on eating habits. This primeval genomic mother mass evolved interactively with events of the social ecology and the physical environment. In such process, *Homo* developed neurobiological mechanisms and systems resulting in an organism of complex individual and collective organization, apt for the development of self-consciousness, search and accumulation of knowledge, symbolic language, technology, and construction of virtual worlds that can be archived and transferable to other generations.

The cultural dimension of the anatomically modern *H. sapiens* in its different phenotypes and ethnic groups has conditioned genetic expression and imprinted a new inheritance added to genomics. We have pointed out in previous chapters the bridge that exists between changes in the hierarchical composition within the group and the state of neuroendocrine regulation, the primary link in the conditioning of social behavior in several species (based on aggressiveness, dominance, submission). Social and environmental ecology are sources of factors that affect neurocognitive and emotional organization and individual behavior.

As stated on another occasion (Colombo, 2013):

> "... we must assume that we are biological entities and therefore belong to the Natural Kingdom, with creative potential and symbolic thought. The amazing thing about it is that such potential has emerged from the evolutionary dynamic -from the existing materials in nature-, in which we must recognize our fossil ancestors."

Implicit in this construction are our cognitive wealth, our limitations and our ambivalences, products of the friction between what has been characterized as *biological and cultural tectonic plates* (Colombo, 2010, 2016).

Through the previous chapters, the development of our species was stressed in a natural evolutionary framework, differentiating it from proposals that try to detach it from the *Natural Kingdom* and make it emerge as an autonomous event, disconnected from natural history. It is true that we have built a macrobiological and expanding culture in which we have taken refuge as the result of that same imperial construction. From it, we engendered pride in our species; a pride that has led us to build an unhealthy relationship with our peers and with ecology to the extreme of placing us at severe risks of future survival and conditioning our creative potential (Colombo, 2010, 2017).

This primordial animal condition that "marks" our fundamental nature came from a prolonged evolutionary existence as *Homo*, which continually interacted with its cultural constructions. It did so in very different ways; sometimes entering into conflict with the limitations imposed by our collective constructions in their attempt to socialize or humanize, or perhaps to domesticate our complex and inventive (creative) nature.

In this socialization/domestication framework, the construction of supernatural agents plays a significant role. Such agents with intentionality were construed from natural phenomena or attributed to inhabitants of the *Natural Kingdom* or virtual beings, or, represented in totemic, iconic, figures (idols, images). This imaginary, virtual universe reinforced by ritual behaviors contributes to control the feeling of helplessness and anxiety caused by phenomena whose nature are not entirely understood or controllable. Additionally, it is also able to generate a feeling of virtual proximity with such virtual agents to which pledges and promises are presented. At the collective level, such rituals generated a feeling of social bonding and shared protection, and a source of familiar promises and prayers; hence, its persistent adherence to collective behaviors.

It is possible to glimpse that nervous circuits associated with the notion of the "acting self" – self-conscience – (Crivelli and Balconi, 2017) overlap to a large extent with those involved in the 'social brain' and the attribution of intentionality to extrapersonal acting agents, as previously commented. As we have seen, these systems are intimately associated with the generation of beliefs and religiosity.

Perhaps the fundamental challenge that our species must face, besides from its survival in a context of virtuous interaction with the social and ecological environment, is to recognize its fundamental animal condition and from there to look for the means to increase its *degrees of creative freedom* (Colombo, 2013) and cooperative social construction. It would be desirable that the invasive technological advances potentially becoming or evolving into *technocracy* and a factor of power do not cancel or replace the emotional potential expressed in the generation and appreciation of beauty and the intrinsic and productive values of solidarity; in short, of those values and behaviors that we would consider characterizing as "humanity".

The classic antinomy between pleasure and duty must abandon the limited definitions that restrain the potency of their meaning and transform through appropriate pedagogical strategies into a synthesis; *the pleasure of merit and duty*. In such sense, pleasure – considered as the resolution of states of tension, and mobilizer of behavior in all orders of the *Natural Kingdom* (Colombo, 2013) – should become one of the central engines of cultural change. Learn to "find" pleasure in merit, solidarity, knowledge, creativity, aesthetic sensibility; in short, in values that optimize human identity from our animal condition. This cultural transformation will depend on whether the *Homo designer* (see below) opts for new "renaissance contents" or for being confused as one more commodity, in the skein of the market rules and their implicit anomie.

10.1. The Designer *Homo*

The construction of "reality" is a dynamic process that consists of an "assemblage" of discrete, unitary elements or images and feelings, according to the species. They fundamentally consist of perceptions and sensations, to which concepts – abstractions – are added in the case of humans. These are based on a neurocognitive substrate configured by effective and relational states and accumulated knowledge. In this sense, the human being can be envisioned as a continuously evolving creature in the role as a "designer" (*Homo quod consilia*, i.e., "the plans of man") in the construction of a vision of "reality".

These final reflections try to support the proposal of a new characterization of *H. sapiens* as *H. sapiens excogitatoris* ("*H. sapiens* designer") or *H. quod consilia* ("*Homo* that plans"), which includes *H. symbolicus* (Henshilwood, 2011), *H. economicus* (S. Mill, 1836, cf. Persky, 1995), or *H. mercatus*, or *H. nundinator* ("merchant") (Colombo, 2010), or *H. faber* vs. *H. luder* and *H. religiosus* (Schiefenhövel, 2009), according to who:

> "The early human mind...needed a powerful paradigm which was able to explain all the seemingly unconnected and unpredictable events of nature as well as its awe-inspiring beauty and functionality, and to make sense of the complexity of human wants and actions. Additionally, our ancestors needed rituals to interact with the supernatural powers who were believed to have created and to be in control of the world. Helplessness, stress, and anxiety were reduced through belief and ritual. It is no wonder that we are, by our very nature, *Homo religiosus*."

As mentioned earlier, a fundamental disagreement should be pointed out regarding the previous statement: *actually, it is not that we intrinsically are H. religiosus, but we culturally became H. religiosus*.

Consider that before the decision of an act (let us say, overt behavior), the individual makes a diagnosis of the situation based on an "experimental" project of ordering the sensed circumstances ("construction of reality" = sensation + perception + memory + cognitive processes). Based on the virtual construction (design) of that reality, it decides how to intervene with its action, be it purely verbal or physical (construction of artifacts, technology).

Therefore, before attributing specific contents or functional characteristics to *Homo* (*sapiens, economicus, faber, ludicus, nundinator, religiosus*), it would be appropriate to define/characterize it by "how it constructs" that reality; before acting, there is a basic preliminary need to have a tentative scheme of "reality" and therefore how to interact with its ecological (social and environmental) framework. What condition lies before the development of instruments (*faber*), the construction of values (*economicus, religiosus, ludicus*) and the construction of knowledge (*sapiens*) based on a given dimension of "reality"? How is it that the fundamental notion of "reality" constructed repeatedly while performing executive behaviors, or during self-contained internal processing, allows us to handle ourselves in all aspects of our daily existence? What is "reality" if not a personal composition, a personal and transient order of a series of limited and interactive perceptions of the external and internal (introspective) universe, and their conceptual connectivity based on previous knowledge and experience? What "data" do we take from those domains? What "frames" do we select to build it? Also, how do we organize them?

According to available dictionaries, the term *sapiens* is equivalent to a possible series of meanings: wise, prudent, sensible, clever. In this sense, its meaning indicates either a condition or a way of acting, or a *Homo* with knowledge indicating that in some way it has acquired and accumulated knowledge and made use of it. Such knowledge can come from the experience for which it had to select the information contents that were appropriate and assemble them in some construction. This is a process where the relationships of all the collected data are *functionally arranged (designed)* into a manageable and instrumental transitory construct of our overt or introspective behaviors. Otherwise, knowledge will arise from tests and counter-tests, for which it will be necessary to exercise the quality of designing and subjecting them to criteria of validity or proof; whether it be conscious or not. Once the data has been obtained/selected, one must, in turn, combine the available options in a useful design and resolve the resulting complexity in a coherent, adaptive, tentative, operative proposal of "reality".

Unlike other species, it is proposed that the essential characteristic in the *Natural Kingdom* that is projected in the *designer or creative Homo* (i.e., *Homo who designs = H. consilia*) consists of a construction based on the limits (capacity) of a *Working Memory* to which we resort in our executive and introspective behaviors which consists of a limited number of events (units) to handle each instance of a behavioral series. These events may be of perceptual, cognitive, affective (emotional), or introspective origin. Eventually, its assembly constructs a "reality" based on a continuous replacement, modulation or reinforcement of such individual events (constituent parts of the "patchwork" with which "reality" is constructed), both in the conscious and subconscious dimensions. In other non-human species, the construction of "reality" would resort to a similar process of "organized patchwork of frames", based on its specific sensory and perceptual characteristics, physiological condition and mnemonic records of the previous experiences. The cognitive domain, the symbolic constructions and the sophistication of the associations (relationships) between events (material or emotional frames) (based on beliefs, assigning values, perspective and retrospective ideation) establish the fundamental difference in the "design" of "reality" of our *Homo* species. Based on the data obtained, we perform a dynamic assembly of those events whether fluid or conditional, which personal or external conditions can modify, and from where we may construct "possible futures".

The "reality" that we construct individually would be nothing more than a sort of multidimensional, conditional, instantaneous, fluid, "puzzle" tinged with variations of the emotional dye based on a personal trait, memory (experience), reflection and conditioned physiological acting. This stored "instantaneous reality", accumulated in mnemonic records, including possible previous errors, will allow the projective, virtual, and anticipatory historical construction of probable future events of "new" realities. Neurocognitive processes relate/associate the unitary components. In this "personal reality", *creativity* and *divergent thinking* will decide the unitary events that we select for such construction and how we relate them to build any given assembly of events. This basic process forms part of adaptation and survival. In order to control for this subjective construction, the scientific method provides a procedure to give a transitory attribute of validity to our cognitive constructions.

The crystallization – loss of fluidity – in our ability to "shuffle" options of the relations between the unitary elements – in conditions of scarcity of elements or cognitive dialectics due to poverty and marginality, or under the pressure of the emotional dimension – will result in a potentially less creative or adaptive condition. *The narrowing of hypothetical options on how to associate unitary events sets up the basis for fundamentalist type of thoughts.*

The dynamic and final content characteristics of such "patchwork" will be affected by the developments imposed by the characteristics of aging and learning during ontogenetic maturation. The pedagogy applied to the *designer* or *creative Homo* will define the dynamics and fluidity of his future designs of "reality". How much awareness is in the field of psycho-pedagogy about the constructive process of the *designer Homo*? Its religious or secular character, its operative evaluation and its objectives, its constructive or destructive intentionality, and its creativity or inventiveness will be defined according to the relationships between its identity and the selected pieces of the "puzzle" based on its cognitive or perceptual, mnemonic and introspective profile.

In this context are particularly interesting the experimental comparative studies (see the previous chapter) that analyze the ascription of agents with intentionality to a set of neutral images and forms that move in a random or organized way, between people assuming the notion of supernatural agents or those that deny them.

Such an individual design of "reality" has a common component inserted through aging, education, rules, and institutions that condition its tendency for the final construction or assembly. That is why we are "designers" of our "outer" and "inner" worlds, conditioned by our nature and cultural relations. Our creativity in such processes is conditioned, and the search for higher *degrees of freedom* is both an individual and a collective endeavor (Colombo, 2013).

This notion of "reality" as a dynamic construction subjected to our nature, our culture and our reflective capacity should be part of our early education. It should provide conditions so that the developing individual "designer" does not see his creativity process prematurely drowned, i.e., blocking the potential maneuvering of relationships among the units of his personal "patchwork". Alternatively, that s/he will not become blind to the fact that the quality of a "graduation as a designer" will depend on a productive – creative – interaction and the *degrees of freedom* (Colombo, 2013, 2017).

It is worth asking then about the ability to designate our species taxonomically as *H. sapiens sapiens* (twice wise, or twice prudent, or sensible, or smart). Based on the previous considerations, it may be appropriate to modify the designation with which we are "tagged" in taxonomy, instead of the wise-wise *Homo* (or prudent-prudent), such as *H. sapiens quod consilia* (*H. sapiens* who plans or designs). As it has been mentioned, the fundamental processes of construction of reality must be standard in the animal kingdom; differences would lay on sensorial and perceptual dynamics of the neurocognitive assembly and the species-specific drives, and the added value based on knowledge and creativity. This phenomenological construction of the maturing *designer Homo* should be pedagogically protected and potentiated during the formative years.

10.2. Beyond Technology: Could We Change the Limits Imposed by our Natural History? How to Deconstruct Inequity

"... inequality is a product of fundamental laws of capitalism ..." (Piketty, 2014) (cf Reeves, 2017).

"The problem of inequality is relational, not economic" (Payne, 2017).

How is the notion of a "different" individual projected toward the social structure of a gregarious community? It relates to the probability of an individual ascending or descending the hierarchy within the said organization. If we take into account the observations made in non-human gregarious communities, the physical characteristics, character, and abilities to establish strategic alliances emerge as differentiating qualities; the right combination results either in the potential figure of the leader or of the subordinate in the social hierarchy, or the marginalized. However, then, there would be doubt about the validity of the claim that "inequality is the product of the fundamental laws of capitalism" (Piketty, 2014) (cf Reeves, 2017).

The antecedents offered by comparative studies on evolutionary social construction and individual positioning in it point to its emergence in gregarious mammalian communities with a hierarchical structure. In other words, in every *non-egalitarian* animal community, conditions are created for inequality of development and individual opportunities. However, it is "*demandable*" that humans manage to generate "utopias" that overcome this conditioning, at least concerning the latter. The individual characteristics should generate the interactive social "mosaic", and not as a product of social deformations that cancel equal opportunities for development.

What type of impact on the neurobiological/neurocognitive and genetic expression on their citizens had each historical forms of sociopolitical organizations? For example, what were their effects on the sociobiological interaction? Beyond the formal differences in their social structure and relations between individuals and institutions, does the engine that feeds them continue to be fueled by dominance, territoriality, predatory cultures, reassurance of a genetic and cultural predominance derived in castes or "groups" – financial, religious, corporate – camouflaged behind the dominant positions?

"Inequity" – or "natural-inequality" expressed in gregarious animal communities with a hierarchical structure – raises an issue of "fundamental animal conditionality" as "naturally inserted" in human social structures. The central question remains on how to bypass such ancestral trend, and to evaluate its impact on the members of the different communities, and on global interaction. Within an anthropological comparative evolutionary approach framework, what mental configuration regarding prosocial behaviors did different political structures or prevailing ideologies favor, that will condition future collective tendencies and sociobiological interactions?

This author doubts whether the present social and political leadership, being mostly focused on short-term problems, is fully aware of the possible future consequences on this *conditionality of the future*; on what individual and collective phenotypes are "baking on the social oven". The plot of economic, financial, religious and ideological interests and their material belongings or interests linked to any of them prevent the leadership from having freedom of action and a clear vision of the consequences of the path undertaken. In that complex web of multitudinous organizations, where is the weight of public opinion, or its expression? Does institutional dynamics – a limited *representative democracy* – degrade and fade out that necessary presence? *Critical knowledge* and *degrees of freedom* are fundamental ingredients to overcome such possibility. Are social and political leaders eager to go through that desirable path? Once again, the nature of early education becomes the key for any substantial progress.

If we turn our attention to the examples provided by the *Natural History* of mammals, and in particular the known history of human societies, it seems that we are facing a dreadful dead end. This picture is construed based on a series of events. Basically, on the role of hierarchies, inequality, annihilation of the opponent or potential threat, cultural predation, violence associated with maladaptation, violence in the imposition of the rules, conditioning of the *degrees of freedom*, privileged minorities, alliances of interests between power factors, demonstration of power with dissuasive purposes, beliefs turned into fundamentalisms. Perhaps with some exceptions of isolated societies, these characteristics seem to have been a constant. At a human level, this uncontrollable sliding of increasing "power-force" manifests itself and perhaps is fully expressed in the recent proposals to generate a "space army force" and to consider future extraterrestrial mining (Lovett, 2012). In the meantime, vast population segments are condemned to material and cognitive impoverishment, or marginalization from knowledge (see Colombo, 2015).

The emergence of humans with a sophisticated language allowed the development of complex virtual constructs based on symbolism and the instruments of culture, which has enhanced cognitive capacity and emotional interaction. However, all of them are supported by processes anchored in neural networks distributed within cortical and subcortical levels, with essential elements preserved during the evolutionary development of the species. As mentioned before, they refer mainly to the hedonic attractiveness of power or prevalence, securing territory, dominance, reproductive and food rights which have become part of the modern cultural process in competition, consumption, accumulation, violence/belligerence, personal contribution, narcissism and privileges – all of these, generators of inequity.

It would seem necessary to mention a previous postulate to the current notions of our evolution as a species: *this did not close the evolution of Homo, and therefore the evolutionary process continues*, but this time from a relatively conscious species of it and with instruments and technology apt to propose different paths or outcomes of the evolutionary process. That is our fundamental responsibility; the present poses ethical, social demands relative to the construction of our collective present and future.

The human, from its origin, has attempted to design their peers through education and their social environment. Today, it has the technology to do it from its genetic base and cultural imposition. What are its limits?

10.3. Considerations on the Scientific-Technological Development and Its Impact on Self-Identity and Cultural Construction

The development of instruments with which to explore the environment and obtain resources by different means recognizes an early presence in anthropoid species that preceded *Homo*. Moreover, its cultural development has depended on the development of instruments with which to interact with the natural elements and their contemporaries, either constructively or destructively. However, in *H. sapiens* it acquired particular significance in the evolutionary process and the organization of communities, as well as in the relative prevalences within them and among them.

With its progressive development, the generation of successive instruments configured a body of scientific-technological knowledge under different power strata of social organization. Its dynamics lay beyond the public domain – i.e., automatization, artificial intelligence, spatial projects – and, hence, exceeds the regulation of its applications and capacity for understanding by a large sector of the world population. At the beginning, fuelled by the incorporation of intra-specific competition, the combination of cooperation and competition boosted technological development to such a point that it has defined the course of political-financial strategies. In the meantime, citizens are bound to incorporate it into their daily life, under media and propaganda pressure, and establish behavioral options for which there was no previous training or meditation, either about its impact on the foundations of social structure – e.g., biomedical options to the issue of genre; consumerism vs. rational needs –, value priorities – e.g., assignment of national budgetary priorities between military and educational/health development; ecological equilibrium, or the relationship between different ethnicities, and cultural Darwinism vs. multicultural promotion. This sort of public "assault" by scientific-technological development and prevalence on cultural idiosyncrasies leads to the result of a high level of an "instrument-dependent" and consumerist culture, to the point of deciding the path of human development.

Perhaps these paragraphs should end with a question mark. Why not prioritize the use of the substantial budgetary resources as a result of corruption, corporate greed, or devoted to military-based projects, applying them to provide more equity in access to essential quality of life and knowledge and ecological survival conditions on the planet? The answer seems to be simple: competition in political and economic interests, and the intrinsic cumulative tendency of those who hold political or financial power appear to have motorized the diversion of funds far beyond the rational and solidary/equity limits.

A few additional considerations should address the unstoppable development of scientific-technological progress, on grounds distant from such goals. This stage in the scientific-technological development poses serious questions regarding its impact on the social organization, as well as on the continuous process of natural selection. If human evolution has to depend on the direct intervention on the genotype and the brain and mental functioning, there is this disturbing question of Harari (2015):

> "When bodies and brains become designer products, will natural selection give way to (technological) intelligent design?"

Who will be the "master designer", or what public cultural awareness level will allow such types of decisions? Will present forms of "representative democracy" under massive lobby interest pressures be enough reassurance of representing public opinion interests and consideration? The interactive dilemma based on an increasing power of corporate and political interests, declining cultural awareness in the educational, and standing of critical information awareness and involvement of the average population poses a dreadful perspective.

Moreover, the messianic behavior of "playing Master" is one of the conceptual redoubts' characteristic of fundamentalisms. It would not be surprising then that the scientific-technological capacity to generate "intelligent designs" of the human being will turn into a new fundamentalist-like battlefield with opponents led by leaders blinded by the desire of dominance – renewing the ancestral animal motivation that does not give way. The tools will be within reach[28].

In summary, postindustrial societies are becoming increasingly dependent on material and technological culture to the point of "embraining" (sic) them. That is, incorporating both of them into basic cognitive daily constructions; conceptually becoming *technological hybrids*. Whether this represents a developmental "must" or an uncontrolled "spin-off" of human inventiveness represents a point that should be taken into immediate conscience at the social and political level.

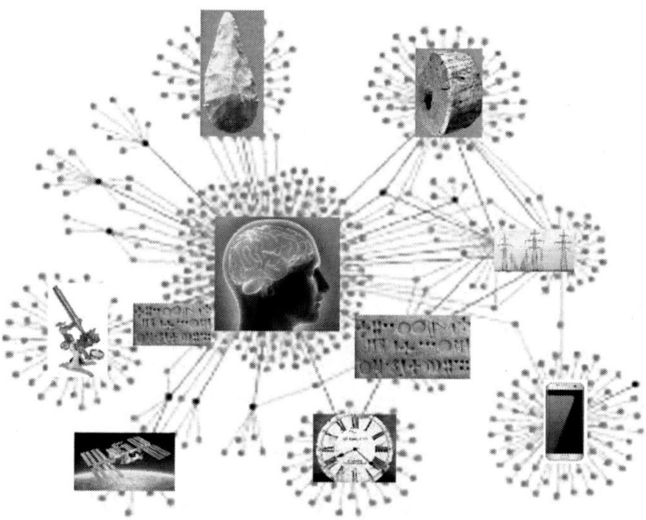

Figure 5. Scheme of the scientific-technological development interactive network generated from the production of successive virtual and material instruments, starting with the arrowhead and the language. (*Author's composite*).

[28] Two examples are given. Genetically editing babies seems to be one recent development, with its positive and also potentially dreadful possibilities, using relatively non-expensive technology for which there already seem to exist offers in the market, according to Wadhwa V. (2018). Also, recent developments on Quantum technology in communication *"Quantum communication is regarded as the most secure because its encryption is based on quantum entanglement, in which two or more subatomic particles affect each other simultaneously, regardless of the distance between them."* (http://www.chinadaily.com.cn/a/201808/28/WS5b84a40ca310add14f388181.html)

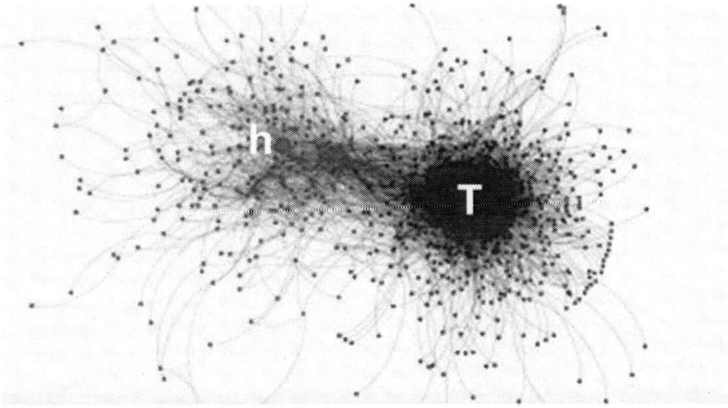

Figure 6. Representation of the dynamics generated by the competence and the dynamics of scientific-technological knowledge (T). Progressive reduction – and replacement – of the dynamics of their development under the control of human communities (h), fuelled by power-seeking and a self-generated developmental dynamic. (Modified, based on social net representation.)

Finally, let us insist on the following global concept expressed on another occasion (Colombo, 2017):

> "Educational public policies that advocate knowledge, the exercise of imagination and divergent thinking or conceptual expansion can only be effective if they are integrated into equity in social construction and the optimization of individual freedom. Indigence, poverty, privilege and class prejudice are its powerful opponents."

SUMMARY ADDENDUM

Figure 7. Our closest, living non-human primate relative. (modified from DLILLC/ Corbis) (https://www.smithsonianmag.com/smart-news/female-chimps-more-likely-males-hunt-tools-180955004/)

A1. TECTONIC PLATES IN THE CONSTRUCTION OF BEHAVIOR

We are biological entities with social history; both events metaphorically expressed as the "BIOLOGICAL-" and the "CULTURAL-TECTONIC PLATES" synthesize the reasons of our desires, frustrations, problems of adaptation, belligerence, aesthetic and social sensitivity.

A2. The Problem of Freedom

Are we aware of the constraints/and conditioning factors to which we are subjected, beyond the particular contexts of each society? The *degrees of freedom* are conditioned by natural biological factors – genetic, physiological, epigenetic – and by historical-cultural contexts. In the latter case, interaction with the environment can generate dysfunctional processes, such as "mood disorders" as a result of the "toxic social factors" involved in imbalances in social adaptation. In this interaction, we modify in the historical evolution several limits that will depend on the individual biological, cultural and socioeconomic characteristics.

A3. The Problem of Coding

Any external signal – norms, estates, constructions of ideas and their internalization – in order to become a cultural heritage and propositional memory, it must have been incorporated into the system of cerebral circuits as readable codes. If it does not open that door, if it is not decodable in biochemical/bioelectrical terms – call it language, symbols, linguistic structures, gestures – it will not have a recordable impact on the construction of behaviors or mental states. As much as they are written in golden letters, it is necessary to enter the mental organization through a suitable organic decoder, reader or receiver. As much as someone could be exposed to the most sensible poems written in Mandarin, it will remain impassive until the meaning of the ideograms is decoded. That is to say, the existence of an external world of norms, rules, relations, signifiers can only be internalized if it complies with that premise: to be configured in such a way as to be able to modify the stationary or dynamic condition of nervous circuits.

A4. How Do We Build Ourselves?

We are built on a genetically driven biological matrix interactive with the environment, with variable margins of individual initiative to which contribute factors related to temperament and personality. They modify each other as long as there are permeable channels and decodable codes. The *degrees of freedom* that condition the creativity potential depend on this multifactorial, dynamic equation. In this sense, the *theory of social impact on brain development* makes of social complexity – *socio-cultural ecology* – a central factor in brain growth, neuroplasticity, and behavioral expression. This theory does not negate the role of epigenetic modulation; environmental modulation is current on the genetic expression of postmitotic cells and heritable expression of DNA by changes in the composition of nuclear chromatin.

A5. How Do We Evolve?

The comments about the fossil evolution of the brain in *Homo* species are based exclusively on indirect morphological data, for the simple reason that at the fossil level it is not possible to detect changes at the molecular level, densities, synaptic properties, or neurotransmitters due to the degradation process. This creates a difficulty – a gap – when trying to correlate those morphological brain changes with the evolution of social complexity among the *Hominids*.

Hence, in order to better understand the implications of such a "gap"on anthropologically evolutive terms, it seems opportune to remark that all construction and transmission of symbolic "artifacts" are necessarily transformed into the relationship between an emitter and a receiver, identifiable by the fact that both expression and reception must be subject to compatible biological coding, as a *sine qua non* condition for the possibility of communication. Operationally, these "social constructs" are indissolubly conditioned to be manipulated by "biological entities with history". Phonemes, words, significant sentences – whether of lesser or greater complexity – must conform to messages decodable by the brain. The expression, reception and interpretation of this "symbolic communication network" will therefore not be alien to the biological condition of the person. The memory of the species is expressed in the acting of being, modified interactively by the individual history and the cultural ecology (physical and emotional).

A6. EXTRABIOLOGICAL ARTIFACTS

Of course, there are cultural "products" that actually exist physically, as "artifacts" in contexts isolated from the biological condition; whether they are written in texts, mentioned in prayers, or projected in design plans, such as a book, a work of art, a musical work, a national constitution, or the Bible. However, their functional existence, as long as they become mental representations, will only be possible insofar as there is a codifying generator of said product and a reader or decoder. Both will be subject to biological and cultural conditions. As mentioned before, the invention of artifacts with "extra biological existence" is characterized by generating products that have their own autonomous existence, once created. However, they will remain isolated, not understood, culturally inert and useless until they are decoded by trained agents to do so. They become "naturalized" once they become functional and ceasing to be ignored artifacts or "objects without a subject". Otherwise, they will maintain their condition of inert constructions stored on dark shelves of an "external memory".

A7. SOCIALIZATION OR DOMESTICATION

No doubt, social interaction is a potential epigenetic modifier, especially in certain early stages of life. Moreover, as each society establishes its own "rules of the game", that society – democratic, autocratic, pluralistic, egalitarian – will try to mold *ab initio* the gold equation, combining influences from home, the school, the pressure from the media and the allowed social interactions. Each society requires a minimum – unequal among them – of obedience and a maximum – also unequal among them – of tolerance to questioning the norms that sustain it.

In this process, the impact on the initial stages of brain/mental development would approach the process of "imprinting" beyond any possibility of critical interaction which suggests the need to distinguish between *socialization* and *domestication.* Later, "education" will contribute conditions for the reinforcement of certain norms, and for the expression of diverse, socially desirable, individual characters that will interact passively or actively with the social structure.

A8. With the Passage of Time, it is Possible to Arrive at the Following Conclusion

It is evident that we must continue struggling to optimize our *degrees of freedom* in terms of operational and thought processes, in order to achieve conditions where socialization and creativity are terms of a desirable equation for the continuous collective development of our talents.

References

Adams, P.J. *Theoretical Biology* 195 (1998): 419-438.

Aiello, L.C., and P.T. Wheeler. *Current Anthropology* 36 (1995): 199-221.

Alcorta, C.S., and R. Sosis. *Human Nature* 16 (2005): 323-359.

Alexander, D. *Genes, Determinism and God,* Cambridge Univ. Press (2017).

Armstrong, E. *Science* 17, 220 (1983): 1302-1304.

Armstrong, E. *Brain Behavior and Evolution* 36 (1990): 166-176.

Atran, S., and J. Henrich. *Biological Theory* 5 (2010): 18-30.

Atran, S., and A. Norenzayan. *Behavioral Brain Sciences* 27 (2004): 713-770.

Attwell, D., and A. Gibb. *Nature Reviews on Neuroscience* 6 (2005): 841-849.

Avants, B.B., P.T. Schoenemann P.T., and J.C. Gee. *Medical Image Analysis* 10 (2006): 397-412.

Axelrod, R., and W.D. Hamilton. *Science* 211 (1981): 1390-1396.

Azari, N.P., J. Nickel, G. Wunderlich, M. Niedeggen, H. Hefter, L. Tellmann, H. Herzog, P. Stoerig, D. Birnbacher, and R.J. Seitz. *European Journal Neuroscience* 13 (2001): 1649-1652.

Azzi, J.C.B., A. Sirigu, and J.R. Duhamel. *Proceedings of the National Academy of Sciences USA* 109 (2012): 2121-2131. http://doi:10.1073/pnas.1111715109.

Ban Ki-moon. https//news.un.org/es/story/2015/06/1332081.

Barks, S.K., L.A. Parr, and J.K. Rilling. *Cerebral Cortex* 25 (2015): 538-544.

Baron-Cohen, S. "Evolution of a theory of mind?" *The Descent of Mind: Psychological Perspectives on Hominid Evolution,* edited by M. Corballis and S. Lea. Oxford University Press, 1999. http://doi: 10.1093/acprof:oso/9780192632593.003.0013.

Barrett, J.L. *Trends in Cognitive Sciences* 4 (2000): 29-34.

Barrett, J.L. *Homo Symbolicus* edited by C.S. Henshilwood and F. d'Errico, Amsterdam/Philadelphia: John Benjamins Publishing Company, Chapter 11, 2011.

Bercik, P., E. Denou, J. Collins, W. Jackson, J. Lu, J. Jury, Y. Deng, P. Blennerhassett, J. Macri, K.D. McCoy, E.F. Verdu, and S.M. Collins. *Gastroenterology* 141 (2011): 599-609.

Berlim, M.T., B.S. Mattevi, P. Belmonte-de-Abreu, and T.J. Crow. *Comprehensive Psychiatry* 44 (2003): 7-14.

Bermúdez de Castro, J.M., J.L. Arsuaga Ferreras, E. Carbonell i Roura, A. Rosas, I. Martínez, and M. Mosquera. "A Hominid from the Lower Pleistocene of Atapuerca, Spain: Possible Ancestor to Neandertals and Modern Humans". *Science* 276 (1997): 1392-1395.

Berridge, K.C., and M.L. Kriegelbach. *Psychopharmacology* 199 (2008): 457-480.

Bissonette, G.B., and M.R. Roesch. *Frontiers in Behavioral Neurosciences* 10 (2016): 1-2.

Blair, R.J. *Brain and Cognition* 55 (2004): 198-208.

Boehm, C. *Science* 336 (2012): 844-847. http//doi:10.1126/science. 1219961.

Boesch, C., and M. Tomasello. *Current Anthropology* 39 (1998): 591-614.

Borre, Y.E., G.W. O'Keeffe, G. Clarke, C. Stanton, T.G. Dinan, and J.F. Cryan. *Trends in Molecular Medicine* September 20(9) (2014): 509-18. http//doi: 10.1016/j.molmed.2014.05.002.

Botero, C.A., B. Gardner, K.R. Kirby, J. Bulbulia, M.C. Gavin, and R.D. Gray. *PNAS* 111 (2014): 16784-16789.

Bouchard, T.J., M. McGue, D. Lykken, and A. Tellegen. *Twin Research Volume 2* (1999): 88-98.

Bouchard, T.J., and M. McGue. *J. Neurobiology* 54 (2003): 4-45.

Boyce, W.T., M.B. Sokolowski, and G.E. Robinson. *Proceedings of the National Academy of Sciences. USA* 109/Suppl. 2 (2012): 17143-17148, http//doi:10.1073/pnas.1121264109.

Boyd, R., and P.J. Richerson. *Philosophical Translations of the Royal Society Lond. B Biol. Sci.* 364 (2009): 3281-3288.

Boyer, P. *Trends in Cognitive Science* 7 (2003): 119-124.

Boyer, P., and P. Lienard. *Behavioral and Brain Sciences* 29 (2006): 1-56.

Bradley, B.J., M.M. Robbins, E.A. Williamson, H.D. Steklis, N.G. Steklis, N. Eckhardt, C. Boesch, and L. Vigilant. *Proceedings of the National Academy of Science. USA* 102 (2005): 9418-9423.

Bravo, J.A., P. Forsythe, M.V. Chew, E. Escaravage, H.M. Savignaca, T.J. Dinan, J. Ienenstock, and J.F. Cryan. *PNAS* 108 (2011): 16050-16055.

Brothers, L. *Concepts in Neuroscience* 1 (1990): 27-51.

Brüne, M. *Neuroscience and Biobehavioral Reviews* 28 (2004): 41-53.

Brüne, M. *Perspectives in Biological Medicine* 49 (2006): 317-329.

Brüne, M. *Philosophy, Ethics and Humanities in Medicine* 2:21 (2007): 1-9. http//doi:10.1186/1747-5341-2-21.

Buchanan, K.L., and D.V. Bohórquez. *Frontiers in Human Neuroscience* 12:323 (2018). http//doi:10.3389/fnhum.2018.00323.

Buffington, S.A., G.V. Di Prisco, T.A. Auchtung, N.J. Ajami, J.F. Petrosino, and M. Costa-Mattioli. *Cell* 165 (2016): 1762-1775. http://doi: 10.1016/j.cell.2016.06.001.

Burkart, J.M., S.B. Hardy, and C.P. Van Schaik *Evolutionary Anthropology* 18 (2009): 175-186.

Burkart, J.M., and P. van Schaik. *Behavioral and Brain Sciences* 39 (2016) http//doi:10.1017/S0140525X15000072, e34.

Burkart, J.M., M.N. Schubiger, and C.P. van Schaik. *Cambridge Journal.* (2016) http://journals.cambridge.org, IP address: 178.195.218.122016.

Burns, J.K. *Behavioral and Brain Sciences* 27 (2004): 831-855.

Burns, J.K. *Progress in Neuropsychopharmacology and Biological Psychiatry* 30 (2006): 797-814.

Burns, J. *The Descent of Madness. – Evolutionary Origins of Psychosis and the Social Brain*, (2007) Routledge, London.

Bykov, K. *La Corteza Cerebral y los Organos Internos* [*The Cerebral Cortex and the Internal Organs*]. Edicion Cartago, 1958.

Bykov, K., G.E. Vladimirov, V.E. Delov, G.P. Konradi, and A.D. Slonim, eds. *Manual de Fisiología* [*Manual of Physiology*], Tomo 3, Editorial Cartago, 1959.

Byrne, R.W., and A. Whiten. *Man New Series* 27 (1992): 609-627.

Byrne, R.W., and N. Corp. *Proceedings of the Royal Society of London* B 27 (2004): 1693-1699.

Carroll, J.B. *Intelligence* 21 (1995): 121-134. http://www.psych.utoronto.ca/users/reingold/courses/intelligence/cache/carroll-gould.html.

Chaimanee, Y., O. Chavasseau, K.C. Beard, A.A. Kyaw, A.N. Soe, C. Sein, V. Lazzari, L. Marivaux, B. Marandat, M. Swe, et al. *Proceedings of the National Academy of Sciences. USA* 109 (2012): 10293-10297.

Chase, I.D., C. Tovey, D. Spangler-Martin, and M. Manfredonia. *PNAS* 99 (2002): 5744-5749.

Chen, C., J. Decety, P.C. Huang, C.Y. Chen, and Y. Cheng. *Human Brain Mapping* May 4 (2016). http//doi:10.1002/hbm.23249.

Chenn, A., and C.A. Walsh. *Science* 297 (2002): 365-369.

Churchill, S.E., and F. Smith. *American Journal of Anthropology* Suppl. 31 (2000): 61-115.

Club of Rome. *The Limits to Growth: A Report for the Club of Rome's Project on the Predicament of Mankind.* Meadows D.H., Meadows D.L., Randers J., Behrens III W.W., Universe Books, New York, 1972.

Cole, S.W. *Current Directions in Psychological Science* 18 (2009): 132-137. http//doi:10.1111/j.1467-8721.2009.01623.

Colombo, E. "La Pobreza Como Categoría Económica" ["Poverty as an Economic Category"] from *Pobreza y Desarrollo Infantil: Uuna Contribución Multidisciplinaria.* (Colombo J.A., ed.), *Paidos*, Buenos Aires, 2007.

Colombo, J.A. *Somos la especie equivocada?* [*Are we the wrong species?*] Ed. EUDEBA, 2010.

Colombo, J.A. *Bajo libertad condicionada* [*Under conditioned freedom*]. Ed. Imago Mundi, Buenos Aires, 2013.

Colombo, J.A. *Los Homo Sabios (globalizados o segmentados?)* [*The Wise Homo (globalized or segmented?)*]. Ed. Buenos Aires Books, 2016.

Colombo, J.A. *Creatividad o domesticación* [*Creativity or domestication*]. Ed. Buenos Aires Books, 2017.

Colombo, J.A. *Pobreza y Desarrollo Infantil* [*Poverty and Child Development*]. (Colombo J.A. editor), Ed. Paidos, Buenos Aires, 2007.

Colombo, J.A., M. Napp, J.R. Depaoli, and V. Puissant. *International Journal of Developmental Neuroscience* 11 (1993): 347-355.

Colombo, J.A. *Brain Structure and Function* Volume 223 (2018): 981-985. htps://doi.org/10.1007/s00429-018-1609-1.

Coqueugniot, H., J.J. Hublin, F. Veillon, F. Houët, and T. Jacob. *Nature* 431 (2004): 299-302.

Cueva, C., R.E. Roberts, T.J. Spencer, N. Rani, M. Tempest, P.N. Tobler, J. Herbert, and A. Rustichini. *Hormones and Behavior* 87 (2017): 1-7.

Curnoe, D., and A. Thorne. *Homo* 53 (2003): 201-224.

Crivelli, D., and M. Balconi. *Frontiers in Behavioral Neuroscience (*2017). http//doi.org10.3389/fnbeh.2017.00229.

Crow, T. *Schizophrenia Research* 28 (1997): 127-141.

Crow, T. *Brain Research Reviews* 31 (2000): 118-129.

Crow, T. *Schizophrenia Research* 102 (2008): 31-52.

Crow, T. *American Journal of Medicine Genetics* Part B 9999 (2013): 1-25.

Cryan, J.F., and T.G. Dinan. *Nature Reviews in Neurosciences* 13 (2012): 701-712.

Darlington, K. *Inteligencia Artificial (3): El Futuro*. [*Artificial Intelligence (3): The Future*]. 2018. https://www. bbvaopenmind.com/ inteligencia-artificial-3-el-futuro/ ?utm_source=newsletter482018Feb&utm_medium=novedadestexto01&utm_campaign=newsletter&utm_content=AIfuture.

Dawkins, R. *The Selfish Gene*, Oxford University Press, Oxford, 1989.

Decety, J. *Annals of the New York Academy of Science* 1231 (2011): 35-42. http//doi.org10.1111/j.1749-6632.2011.06027.

Decety, J., G.J. Norman, G.G. Berntso, and J.T. Cacioppo. *Progress in Neurobiology* 98 (2012): 38-48.

Decety, J., and J.M. Cowell. *Perspectives on Psychological Science* 9 (2014): 525-537.

Decety, J., I.B. Bartal, F. Uzefovsky, and A. Knafo-Noam. *Philosophical Translations of the Royal Society of London B Biol. Sci.* Jan (2016): 19-371. (1686): 20150077. http//doi: 10.1098/rstb.2015.0077.

De Dreu, C.K.W., L.L. Greer, G.A. Van Kleef, S. Shalvi, and M.J.J. Handgraaf. *PNAS* 108 (2011): 1262-1266.

Deeley, P.Q. *J. Anthropology and Medicine* 11 (2010): 245-267.

Dein, S., and R. Littlewood. *Transcultural Psychiatry* 48 (2011): 318-335.

Demertzi, A., A. Van Ombergen, E. Tomilovskaya, B. Jeurissen, E. Pechenkova, C. Di Perri, L. Litvinova, E. Amico, A. Rumshiskaya., I. Rukavishnikov, et al. *Brain Structure and Function* 221 (2016): 2873-2876.

Dennet, D.C. *Breaking the Spell: Religion as a Natural Phenomenon*. London, England: Penguin Books, 2006.

Desrosiers, R., K. Frierici, and F. Rottman. *Proceedings of the National Academy of Science* USA 71 (1974): 3971-3975.

De Waal, F. *Our Inner Ape*. New York: Berkley Publishing Group, The Penguin Group, 2005.

De Waal, F. *Chimpanzee Politics*. Baltimore, Maryland: The Johns Hopkins University Press, 2007.

Diaz Heijtz, R., S. Wang, F. Anuar, Y. Qian, B. Björkholm, A. Samuelsson M.L. Hibberd, H. Forssberg, and S. Pettersson. *PNAS* 108 (2011): 3047-3052.

Dietrich, A. *How Creativity Happens in the Brain*. New York: Palgrave-Macmillan, 2015.

Dominissini, D., S. Nachtergaele, S. Moshitch-Moshkovitz, E. Peer, N. Kol, M.S. Ben-Haim, Q. Dai, A. Di Segni, M. Salmon-Divon, W.C. Clark, et al. *Nature* 530 (2016): 441-446, http//doi.org.10.1038/nature16998.

D'Onofrio, B.M., L.J. Eaves, L. Murrelle, H.H. Maes, and B. Spilka. *Journal of Personality* 67 (1999): 953-984.

Dorus, S., E.J. Vallender, P.D. Evans, J.R. Anderson, S.L. Gilbert, M. Mahowald, G.J. Wyckoff, C.M. Malcom, and B.T. Lahn. *Cell* 119 (2004): 1027-1040.

Dunbar, R.I.M. *Behav. Brain Sci.*16 (1998): 681-735.

Dunbar, R.I.M. *Evolutionary Anthropology* 6 (1998): 178-190.

Dunbar, R.I.M. *Approaches to the Evolution of Language*. Hurford J.R., Studdert-Kennedy M., Knight C., eds. Cambridge: Cambridge University Press, (1999): 92-110.

Dunbar, R.I.M. "Evolutionary Basis of the Social Brain," *The Oxford Handbook of Social Neuroscience.* Cap. 3 (Decety J., Cacioppo J.T, eds.), Oxford Univ. Press, 2011.

Dunbar, R.I.M. *Annual Review of Anthropology* 32 (2003): 163-181.

Dunbar, R.I.M. *Annals of Human Biology* 36 (2009): 562-572.

Dunbar, R.I.M., and S. Shultz. *Philosophical Transactions of the Royal Society* (B) 362 (2007): 649-658. http//doi:10.1098/rstb.2006.2001.

Dunbar, R.I.M. and S. Shultz. *Science* 317 (2007): 1344-1347.

Durkheim, E. *The Elementary Forms of Religious Life*, New York: The Free Press, 1995.

Editorial. *Nature* 534 (2016): 151.

EFE. *A Pedradas Contra el Diablo*. [*Stones against the Devil*]. https://elpais.com/internacional/2009/11/27/actualidad/1259276412_850215.html.

Eilam, D. *Neuroscience and Biobehavioral Reviews* 49 (2015): 55-70.

Eilam, D., R. Izhar, and J. Mort. *Neuroscience and Biobehavioral Reviews.* (2010). doi: 10.1016/j.neubiorev.2010.08.002.

Eliade, M. *Shamanism Archaic Techniques of Ecstasy*. Princeton: Bollingen Foundation, Princeton University Press, 1964.

Ellen, R., and M.D. Fischer. *Understanding Cultural Transmission In Anthropology: A Critical Synthesis,* (Ellen R., Lycett S.J., Johns S.E. eds.). Berghahn Books, N.Y., 2013.

Elliott, R., K.J. Friston, and R.J. Dolan. *Journal of Neuroscience* 20 (2000): 6159-6165.

Ellul, J. *The Technological Society*. New York: Vintage book, A.A. Knopf, Inc. and Random House Inc., 1964.

Elvis, M. *Nature* 485 (2012): 549.

Enard, W., M. Przeworski, S.E. Fisher, C.S. Lai, V. Wiebe, T. Kitano, A.P. Monaco, and S. Pääbo. *Nature* 418 (2002): 869-872.

Evans, P.D., J.R. Anderson, E.J. Vallender, S.L. Gilbert, C.M. Malcom, S. Dorus, and B.T. Lahn. *Human and Molecular Genetics* 13 (2004): 489-494.

Fabricius, E. *La Conducta de Los Animales* [*The Conduct of Animals*]. Buenos Aires: EUDEBA, 1966.

Fauth, M., and C. Tetzlaff. *Frontiers in Neuroanatomy* Vol.10 (2016). http//doi:10.3389/fnana.2016.00075.

Fernald, R.D., and K.P. Maruska. *Proceedings of the National Academy of Science USA* 109, Suppl. 2 (2012):17194-17199, 2012. doi: 10.1073/pnas.1202552109.

Fischer, R., R. Callandar, P. Reddish, and J. Bulbulia. *Journal of Human Nature* 24 (2013): 115-125. doi: 10.1007/s12110-013-9167-y.

Fiske, A.P., and N. Haslam. *Journal of Nervous and Mental Disease* 185 (1997): 211-222.

Flannelly, K.J., H.G. Koenig, K. Galek, and C.G. Ellison. "Beliefs, Mental Health and Evolutionary Threat Assessment Systems in the Brain." *Journal of Nervous and Mental Disease* 195 (2007): 996-1003.

Flinn, M.V., D.C. Geary, and C.V. Ward. *Evolution of Human Behavior* 26 (2005): 10-46.

Foley, R., and C. Gamble. *Philosophical Transactions of the Royal Society of London B Biol. Sci.* 364 (2009): 3267-3279.

Foley, R., L. Martin, M.M. Lahr, and C. Stringer. *Philosophical Transactions of the Royal Society of London*. B 371 (2016): 1-8. http://dx.doi.org/10.1098/rstb.2015.0229.

Fonseca-Azevedo, K., and S. Herculano-Houzel. *PNAS* 109 (2012): 18571-18576.

Forterre, P. *PNAS* 103 (2006): 3669-3674.

Forterre, P. *Studies in History and Philosophy of Biological and Biomedical Sciences* 59 (2016): 100-108. doi: 10.1016/j.shpsc.2016.02.013.

Fumagalli, M., and A. Priori. *Brain* 135 (2012): 2006-2021.

Furuichi, T. *Evolutionary Anthropology* Jul-Aug; 20/4 (2011): 131-142. http//doi:10.1002/evan.20308.

Gagneux, P., and A. Varki. *Molecular and Phylogenetic Evolution* Jan;18/1 (2001): 2-13.

Galef, B., Jr., *Human Nature.* New York: Walter de Gruyter (ed.), Inc., 3 (1992): 157-178.

Garrido, O.G. *Tirales Piedras al Diablo* [*Throwing Stones at the Devil*]. https://elpais.com/elpais/2018/08/23/opinion/1535033429_635319.html.

Gavrilets, S., and A. Vose. *Proceedings of the National Academy of Science USA* 103: (2006): 16823-16828.

Ghiglieri, M. P. *Journal of Human Evolution* (1987): 19-357.

Ghirlanda, S., S. Lind, and M. Enquist. *Royal Society Open Science* 4 (2017): 161011. http://dx.doi.org/10.1098/rsos.161011.

Glansdorff, N. *Molecular Microbiology* 38 (2000): 177-185.

Glansdorff, N., Y. Xu, and B. Labedan. *Biology Direct* 3 (2008): 1-35. http//doi:10.1186/1745-6150-3-29.

Glazier, D.S. *Biological Reviews* 80 (2005): 611-662.

Glazier, D.S. *Biological Reviews* 85 (2010): 111-138.

Gobrogge, K.L., Y. Liu, X. Jia, and Z. Wang. *Journal of Comparative Neurology* 502 (2007): 1109-1122.

Gordon, T.P., R.M. Rose, C.L. Grady, and I.S. Berstein. *Folia Primatologica* (Basel). 32 (1979): 149-160.

Gould, S.J. *The Mismeasure of Man.* New York: W.W. Norton & Co., 1981.

Gräff, J., and L.H. Tsai. *Nature Reviews on Neuroscience* 14 (2013): 97-111. doi:10.1038/nrn3427.

Graybiel, A.M. *Annual Review on Neuroscience* 31 (2008): 359-387. doi: 10.1146/annurev.neuro.29.051605.112851.

Gruber, T., and Z. Clay. *Evolutionary Anthropology* 25 (2016): 239-252.

Haber, S.N., and B. Knutson. *Neuropsychopharmacology Rev.* 35 (2010): 4-26.

Harari, Y.N. *De Animales a Dioses* [*From Animals to Gods*], Bogota, Colombia: Penguin Random House, 2015.

Harari, Y.N. *Homo Deus. A Brief History of Tomorrow*. Digital Editorial: Titivillus ePub base r1.2, 2015.

Hare, B., J. Call, B. Agnetta, and M. Tomasello. *Animal Behaviour,* 59 (2000): 771-785.

Hare, B., and M. Tomasello. *Animal Behaviour* 68 (2004): 57-581. doi: 10.1016/j.anbehav.2003.11.011.

Hare, B., J. Call, and M. Tomasello. *Cognition* 101 (2006): 495-514.

Harlow, H.F. *Scientific American* 200 (1959): 68,70,72-73,74.

Harris, S., J.T. Kaplan, A. Curiel, S.Y. Bookheims, M. Iacoboni, and M.S. Cohen. *PLoS ONE* 4(10): e0007272. doi: 10.1371/journal.pone. 0007272.

Hayden, B.Y., D.V. Smith, and M.L. Platt. *Proceedings of the National Academy of Science USA* 106 (2009): 5948-5953.

Heard, E., and R.A. Martienssen. *Cell* 157 (2014): 95-109. doi: 10.1016/j.cell.2014.02.045

Hebb, D.O. *The Organization of Behavior. A Neuropsychological Theory*. Londres: John Wiley & Sons, 1949.

Henrich, J., and C. Tennie. https://www2.psych.ubc.ca/~henrich/pdfs/ Henrich_Tennie_2015_revised_final_final.pdf.

Henshilwood, C.S. *Homo Symbolicus*, Amsterdam/Philadelphia: John Benjamins Publishing Company, 2011.

Hodder, I. *New Literary History* 45 (2014): 19-36.

Horner, V., and F.B.M. de Waal. *Progress in Brain Research*, Chiao J.Y. (ed.), 178 (2009): 3-15.

Horrobin, D.F. *Medical Hypothesis.* 50 (1998): 269-288.

Howlett, J.R., and M.P. Paulus. *PLos ONE* 10 (2015): 1-17. e0124596. doi:10.1371/journal. pone.0124596.

Ikemoto, S., and J. Panksepp. *Brain Research Reviews* 31 (1999): 6-41.

Instituto Médico Howard Hughes (HHMI). *Comparación de Los Genomas de Las Especies* [*Comparison of Species Genomes*], 2009. http://www. actionbioscience.org/esp/genomica/ hhmi.html.

Inzlicht, M., A.M. Tullet, and M. Good. *Religion Brain and Behaviour* 1 (2012): 192-212. doi.org/10.1080/2153599X.2011.647849.

Jablonka, E., and M.J. Lamb. *Behavioral and Brain Sciences* 30 (2007): 353-365.

Jacques, S. *Neurosurgery* 5 (1979): 277-283.

Jakob, C. *Folia Neurobiológica Argentina* [*Foil Neurobiology Argentina*], Tomo 1 Neurobiología General, Buenos Aires: As Aniceto López, 1941.

Jensen, A.R. *Contemporary Education Rev.*1 (1982): 121-135. https://www.debunker.com/texts/jensen.html.

Jia, G., Y. Fu, X. Zhao, Q. Dai, G. Zheng, Y. Yang, C. Yi, T. Lindahl, T. Pan, Y.G. Yang, and C. He. *Nature Chemical Biology* 7 (2011): 885-887. doi:10.1038/nchembio.687.

Kaessmann, H., and S. Pääbo. *Journal of Internal Medicine* 251 (2002): 1-18.

Kapogiannis, D., A.K. Barbeya, M. Su, G. Zamboni, F. Krueger, and J. Grafman. *PNAS* 106 (2009): 4876-4881.

Kapogiannis, D., G. Deshpande, F. Krueger, M.P. Thornburg, and J.H. Grafman. *Brain Connectivity* 4 (2014): 70-78.

Karlsson, S.A., E. Studer, P. Kettunen., and L. Westberg. *Frontiers in Behavioral Neuroscience* (2016). http://dx.doi.org/10.3389/fnbeh.2016.00041.

Kauffman, P.R. *Journal of Nervous and Mental Disease* 204 (2016): 702-712.

Kehrer-Sawatzki, H., and D.N. Cooper. *Human Genetics* 120 (2007): 759-778.

Kerhoas, D., D. Perwitasari-Farajallah, M. Agil, A. Widdig, and A. Engelhardt. *Behavioral Ecology* 25 (2014): 1164 – 1172. doi:10.1093/beheco/aru099.

Kitayama, S., and A.K. Uskul. *Annual Review on Psychology* 62 (2011): 419-449.

Kleiber, M. *Physiological Review* 27 (1947): 511-541.

Knutson, B., C.M. Adams, G.W. Fong, and D. Hommer. *Journal of Neuroscience* 21: RC159 (2001): 1-51.

Koenig, L.B., M. McGue, and T.J. Bouchard, Jr. *Journal of Person* 73 (2005): 471-488.

Koenig, L.B., M. McGue, and W.G. Iacono. *Developmental Psychology* 44 (2008): 532-543. doi:10.1037/0012-1649.44.2.532.

Koenig, L.B., M. McGue, and W.G. Iacono. *Personality and Individual Differences* 47 (2009): 652-656. doi: 10.1016/j.paid.2009.06.003.

Kohler, T.A, M.E. Smith, A. Bogaard, G.M. Feinman, C.E. Peterson, A. Betzenhauser, M. Pailes, E.C. Stone, A.M. Prentiss, T.J. Dennehy, et al. *Nature*. doi: 10.1038/nature24646.

Kojima, T., H. Onoe, K. Hikosaka, K. Tsutsui, H. Tsukada, and M. Watanabe. *Journal of Neuroscience* 29 (2009): 14463-14471.

Konvalinka, I., D. Xygalatas, J. Bulbulia, U. Schjødt, E.M. Jegindø, S. Wallot, G. Van Orden, and A. Roepstorff. *PNAS* 108 (2011): 8514-8519.

Kouprina, N., A. Pavlicek, G.H. Mochida, G. Solomon, W. Gersch, Y-H. Yoon, R. Collura, M. Ruvolo, J.C. Barrett, and CG. Woods. *PLoS Biology* (2004): 0653-0663. http://biology.plosjournals.org 2.

Krech, D., M.R. Rosenzweig, and E.L. Bennett. *Journal of Comparative and Physiological Psychology* 53 (1960): 509-514.

Kuhl, H.S., A.K. Kalan, M. Arandjelovic. et al. *Nature Science Reports* 6 (2016): 22219. doi: 10.1038/srep22219 (2016).

Kuhlwilm, M., I. Gronau, M.J. Hubisz, C. de Filippo, J. Prado Martinez, M. Kircher, Q. Fu, H.A. Burbano, C. Lalueza-Fox, et al. *Nature* 530 (2016): 429-433. doi: 10.1038/nature16544.

Lahr, M.M. *Philosophical Transactions of the Royal Society* B 371 (2016): 1-12. http://dx.doi.org/10.1098/rstb.2015.0241.

Lamm, C., J. Decety, and T. Singer. *Neuroimage* 54 (2011): 2492-2552. doi: 10.1016/j.neuroimage.2010.10.014.

Lang, M., J. Krátký, J.H. Shaver, D. Jerotijević, and D. Xygalatas. *Current Biology* 25 (2015): 1892-1897. doi: 10.1016/j.cub.2015.05.049.

Latzman, R.D., L.K. Hecht, H.D. Freeman, S.J. Schapiro, and W.D. Hopkins. *Neuroimage* 123 (2015): 63-71.

LeDoux, J. *Neuron* 73 (2012): 653-676.

Leonard, W.R., M.L. Robertson, J.J. Snodgrass, and C.W. Kuzawa. *Comparative Biochemistry and Physiology Part A* 136 (2003): 5-15.

Leroi-Gourhan, A. "The Flowers Found with Shanidar IV, a Neanderthal Burial in Iraq." *Science* 190 (1975): 562-564.

Lewis, G.J., and T.C. Bates. *Journal of Research in Personality* 47 (2013): 398-405.

Lewis-Williams, J.D. *Belief in the Past: Theoretical Approaches to the Archaeology of Religion.* New York: Whitley D.S., Hays-Gilpin Routledge K. (eds.) 2016.

Lewitus, E., I. Kelava, A.T. Kalinka, P. Tomancak, and W.B. Huttner. *PLoS Biol.* 12 (2014): e1002000, doi: 10.1371/journal.pbio.1002000.

Ley, R.E., C.A. Lozupone, M. Hamady, R. Knight, and J.I. Gordon. *Nature Review on Microbiology* 6 (2008): 776-788.

Lind, J., P. Lindenfors, S. Ghirlanda, K. Liden, and M. Enquist. *Scientific Reports 3*: 1785. doi: 10.1038/srep01785.

Lind, J., S. Lönnberg, T. Persson, and M. Enquist. *Frontiers in Psychology* 8 (2017): 161. doi:10.3389/fpsyg.2017.00161. 2017.

Lipina, S.J., and J.A. Colombo. "Poverty and Brain Development during Childhood." *American Psychological Association*, Washington DC, 2009.

Lipina, S.J., M.S. Segretin, M.J. Hermida, and J.A Colombo. "Research on Childhood Poverty from a Cognitive Neuroscience Perspective: Examples of Studies in Argentina." *International Handbook on Mental Health of Children and Adolescents: Culture, Policy & Practices.* Delhi, India. SAGE, 2012.

Liu, F., W. Clark, G. Luo, X. Wang, Y. Fu, J. Wei, X. Wang, Z. Hao, Q. Dai, and G. Zheng et al. *Cell* 167 (2016): 816-828, 2016. doi: 10.1016/j.cell.2016.09.038.

Lorenz, K.Z. *Philosophical Transactions of the Royal Society* B 251 (1966): 273-284.

Lovett, R.A. *Nature* 2012, doi:10.1038/nature.2012.10513.

Lubin, F.D., S. Gupta, S. Parrish, R.R., N.M. Grissom, and R. L. Davis. *Neuroscientist* Dec. 17(6) (2011): 616-32. doi: 10.1177/107385 8411386967.

Luo, G-Z., M.A. Blanco, E.L. Greer, C. He, and Y. Shi. *Nature Review on Molecular Biology* 16 (2015): 705-710. doi:10.1038/nrm4076.

Lykken, D.T., T.J. Bouchard, Jr., M. McGue, and T. Auke. *J. Applied Psychology* 78 (1993): 649-661.

MacLean, P.D. *The Triune Brain in Evolution*. New York: Plenum Press, 1990.

MacLean, E.L. *PNAS* 113 (2016): 6348-6354.

Martos, Y.V., B.Y. Braz, J.P. Beccaria, M.G. Murer, and J.E. Belforte. *Journal of Neuroscience.* 37 (2017): 2849-2858.

Maruska, K.P., Zhang A., Neboori A., and Fernald R.D. *Journal of Neuroendocrinology* 25 (2013): 145-157. doi: 10.1111/j.1365-2826.2012.02382.x.

Mauer, J., X. Luo, A. Blanjoie, X. Jiao, A.V. Grozhik, D.P. Patil, B. Linder, B.F. Pickering, J.J. Vasseur, and Q. Chen. *Nature* 541 (2017): 371-375, doi:10.1038/nature21022.

Mauss, M. *A General Theory of Magic*, translated and published by Robert Brain, London, 1972, from the original in *Anné Sociologique*, 1902.

Mcbrearty, S., and A.S. Brooks. *Journal of Human Evolution* 39 (2000): 453-563.

McCauley, R.N., and E.T. Lawson. *Bringing Ritual to Mind: Psychological Foundations of Cultural Forms*. Cambridge and New York: Cambridge University Press, 2002.

McClure, S.M., M.K. York, and P.R. Montague. *Neuroscientist* 10 (2004): 260-268.

McConkey and Goodman. (1997), see in Kehrer-Sawatzki et al. 2007.

McGowan, P.O., and M. Szyf. *Neurobiology of Disease* 39 (2010a): 66-72.

McGowan, P.O., and M. Szyf. *Essays in Biochemistry* 48 (2010b): 275-287. doi:10.1042/bse0480275.

McGrew, W.C. *Human Ethology Bulletin, Proc. of the XXII. ISHE Conference* (2015): 41-52.

McKenna, T. *Food of the Gods*, NY: Bantam New Age Books, 1992.

Meaney, M.J., and A.C. Ferguson-Smith. *Nature Neuroscience* 13 (2010): 1313-1318. doi:10.1038/nn1110-1313.

Meloni, M. *Frontiers in Human Neuroscience.* www.frontiersin.org 8 (2014): 1-12.

Milner, P.M. *Canadian Journal of Psychology* 45 (1991): 1-36.

Mithen, S. *The Prehistory of the Mind: A Search for the Origins of Art, Religion, and Science*. London: Thames and Hudson Ltd, 1996.

Moll, H., and M. Tomasello. *Philosophical Transactions of the Royal Society* (B) 362 (2007): 639-648. doi: 10.1098/rstb.2006.2000.

Moll, J., R. de Oliveira-Souza, I.E. Bramati, and J. Grafman. *Neuroimage* 16 (2002): 696-703.

Montoya, E.R., D. Terburg, P.A. Bos, G.J. Will, V. Buskens, W. Raub, and J. van Honk. *Psychoneuroendocrinology* 38 (2013): 1362-1369. doi: 10.1016/j.psyneuen.2012.12.001.

Navarrete, A., C.P. van Schaik, and K. Isler. *Nature* 480 (2011): 91-93. doi: 10.1038/nature10629.

Neubauer, R.L. *Brain & Behavior* 4 (2014): 92-103.

Noonan, J.P., G. Coop, S. Kudaravalli, D. Smith, J. Krause, J. Alessi, F. Chen, D. Platt, S. Pääbo, and J.K. Pritchard. *Science* 314 (2006): 1113-1118.

Norenzayan, A., and A.F. Shariff. *Science* 322 (2008): 58-62.

Norenzayan, A., and W.M. Gervais. *Trends in Cognitive Sciences*. 17 (2013): 20-25.

Obayashi, S., Y. Nagai, T. Suhara, T. Okauchi, M. Inaji, A. Iriki, and J. Maeda. *Neuroscience Research* 64 (2009): 421-428.

O'Connell, L.A., and H.A. Hofmann. *Front. Neuroendocrinol.* 32 (2011): 320-335. doi: 10.1016/j.yfrne.2010.12.004.

Olds, J., and P.M. Milner. *Comparative Physiological Psychology* 47 (1954): 419-427.

Oliveira, R.F., and J. Fish. *Biology* 81 (2012): 2127-2150. doi: 10.1111/j.1095-8649.2012.03477.

Osada, N., and C.I. Wu. *Genetics* 169 (2001): 259-264.

Panksepp, J. *Conscious Cognition* 14 (2005): 30-80.

Pascual, L., P. Rodrigues, and D. Gallardo-Pujol. *Frontiers in Integrative Neuroscience* 7 (2013): 65. doi: 10.3389/fnint.2013.00065.

Patterson, N., D.J. Richter, S. Gnerre, E.S. Lander, and D. Reich. *Nature* 441 (2006): 1103-1108.

Payne, K. *The Broken Ladder*. New York: Penguin Random House, 2017. http:// www.penguinrandomhouse.com/ books/ 318941/ the-broken-ladder-by-keith-payne/9780525429814/.

Peirce, C.S., and M. Tomasello. *The Cultural Origins of Human Cognition*. Harvard University Press, 1999. https://es.scribd.com/document/ 53978890/The-Cultural-Origins-of-Human-Cognition-1999.

Pennycook, G., R.M. Ross, D.J. Koehler, and J.A. Fugelsang. *PLoS ONE* 11(4): e0153039. doi: 10.1371/journal.pone.0153039.

Peoples, H.C., and F.W. Marlowe. *Human Nature* 23 (2012): 253-269. doi: 10.1007/s12110-012-9148-6.

Peoples, H.C., P. Duda, and F.W. Marlowe. *Human Nature* 27 (2016): 261-282. doi: 10.1007/s12110-016-9260-0.

Persinger M.A. *Journal of Neuropsychiatry and Clinical Neuroscience* 13 (2001): 515-524.

Persky, J. "Retrospectives: The Ethology of Homo Economicus." *The Journal of Economic Perspectives* 9 (1995): 221-231. https://es. scribd.com/document/350459027/jep-9-2-221. https://es.wikipedia.org/ wiki/Homo_%C5%93conomicus.

Pessiglione, M., L. Schmidt, B. Draganski, R. Kalisch, H. Lau, R.J. Dolan, and C.D. Frith. *Science* 316 (2007): 904-906.

Piaget, J. "La Causalité Chez L'Enfant" ["Causality in children"]. *British J. Psychol.* doi:10.1111/ j.2044-8295.1928.tb00466.x.

Piketty, T. (2014), see in Reeves A., 2017.

Pinto-Sánchez, M., G.B. Hall, K. Ghajar, C. Bolino, J.T. Lau, F.P. Martin, et al. *Gastroenterology* Aug;153(2) (2017): 448-459. e8. doi: 10.1053/j.gastro.2017.05.003.

Polimeni, J., and J.P. Reiss. *Medical Hypotheses* 58 (2002): 244-248.

Poutahidis, T., S.M. Kearney, T. Levkovich, P. Qi, B.J. Varian, J.R. Lakritz, Y.M. Ibrahim, A. Chatzigiagkos, E.J. Alm, and S.E. Erdman. *PLoS ONE* 8/10 (2013). e78898. doi: 10.1371/journal.pone.0078898.

Pradhan, G.R., C. Tennie, and C.P. van Schaik. *Journal of Human Evolution* 63 (2012): 180-190.

Preston, S.D. *Psychological Bulletin* 139 (2013): 1305-1341. doi: 10.1037/a00317552013.

Previc, F.H. *Brain and Cognition* 41 (1999): 299-355.

Previc, F.H. *Conscious Cognition* 15 (2006): 500-539.

Previc, F.H. *Dopaminergic Mind in Human Evolution and History*. UK: Cambridge University Press, 2009.

Purzycki, B.G., J. Henrich, C. Apicella, Q.D. Atkinson, A. Baimel, E. Cohen, R.A. McNamara, A.K. Willard, D. Xygalatas, and A. Norenzayan. *Religion, Brain & Behavior* (2017). http://dx.doi.org/10.1080/2153599X.2016.1267027.religion.

Raichle, M. *Science* 314 (2006): 1249-1250.

Raichle, M., A. MacLeod, A. Snyder, W. Powers, D. Gusnard, and G. Shulman. *Proceed. Natl. Acad. Sci. USA* 98 (2001): 676-682.

Rakic, P. *Trends Neurosci.* 18 (1995a): 383-388.

Rakic, P. *Proc. Natl. Acad. Sci.* USA 92 (1995b): 11323-11327.

Rash, B.G., J.B. Ackman, and P. Rakic. *Science Advances 2*: e1501733,26 February 2016.

Reader, S.M., and K.N. Laland. *PNAS* 99 (2002): 4436-4441.

Reader, S.M., Y. Hager, and K.N. Laland. *Philosophical Transactions of the Royal Society* (B) 366 (2011): 1011-1027. doi:10.1098/rstb.2010.0342.

Reeves, A. "Economics: the Architecture of Inequality." *Nature* 543 (2017): 312-313.

Renfrew, C. "The Archaeology of Religion." *The Ancient Mind.* (Renfrew C., Zubrow E., eds.) Chapt. 6, Cambridge University Press, 1994.

Renfrew, C. *Proceedings of the British Academy*.139 (2006): 217-238.

Renfrew, C. *Philosophical Transactions of the Royal Society* B 363 (2008): 2041-2047. doi:10.1098/rstb.2008.0010.

Rieder, R., P.J. Wisniewski, B.L. Alderman, and S.C. Campbell. *Brain, Behavior, and Immunity* (2017). DOI: http://dx.doi.org/10.1016/j.bbi.2017.01.016.

Riekki, T., M. Lindeman, and T.T. Raij. *Journal of Social Neuroscience* 9 (2014): 400-411.

Rilling. J.K. and T.R. Insel. *Neuroreport* 10 (1999): 1453-1459.

Rilling, J.K. J. Scholz, T.M. Preuss, M.F. Glasser, B.K. Errangi, and T.E. Behrens. *SCAN* 7 (2012): 369-379. doi: 10.1093/scan/nsr017.

Rilling, J.K. *Trends in Cognitive Sciences* 18 (2014): 46-55.

Ringo, J.L. *Brain Behav. Evol.* 38 (1991): 1-6.

Robbins, M.M. *Anim. Behav*. 57 (1999): 1013-1020.

Robertson, D.R. *Science* 177 (1972): 1007-1009.

Robertson, D.M., R. Paganelli, R. Dinwiddie, and R.J. Levinsky. *Archives of Disease in Childhood* 57 (1982): 369-372.

Robinson, G.E., C.M. Grozinger, and C.W. Whitfield. *Nature Reviews on Genetics* 6 (2005): 1-15.

Robinson, G.E., R.E. Fernald, and D.F. Clayton. *Science* 322 (2008): 896 – 900. doi: 10.1126/science.1159277.

Rose, R.M., T.P. Gordon, and I.S. Bernstein. *Science* 178 (1972): 643-645.

Rossano, M.J. *Review of General Psychology American Psychol.* Ass. 10 (2006): 346 – 364. doi: 10.1037/1089-2680.10.4.346.

Rossano, M.J. *The Biological Evolution of Religious Mind and Behavior* (Eds. Voland E., Schiefenhövel W.), Chap. 9 (2009): 127-143. Elsevier, Heidelberg.

Roth, G., and U. Dicke. *Trends in Cognitive Science* 9 (2005): 250-257.

Rudenko, A., and L.H. Tsai. *Neuropharmacology* 80 (2014): 70-82. doi: 10.1016/j.neuropharm.2014.01.043.

Sagan, C. *El Mundo y Sus Demonios* [*The world and its demons*]. Barcelona, Spain: Editorial Planeta, 1995.

Sallet, J., R.B. Mars, M.P. Noonan, J.L. Andersson, J.X. O'Reilly, S. Jbabdi, P.L. Croxso, M. Jenkinson, K.L. Miller, and M.F.S. Rushworth. *Science* 334 (2011): 697-700.

Santos, L.R., and A.G. Rosati. *Annual Review in Psychology* 66 (2015): 321-347. doi: 10.1146/annurev-psych-010814-015310.

Schiefenhövel, W. *The Biological Evolution of Religious Mind and Behavior*. (Voland E, Schiefenhövel W., eds.), Chap. 10, Springer-Verlag, Berlin, 2009.

Schjoedt, U., H. Stødkilde-Jørgensen, A.W. Geertz, and A. Roepstorff. doi: 10.1093/scan/nsn050 SCAN 4 (2009): 199-207.

Schjoedt, U., J. Sørensen, K.L. Nielbo, D. Xygalatas, P. Mitkidis, and J. Bulbulia. *Religion, Brain and Behavior* 3 (2013): 39-55.

Schloss, J.P., and M.J. Murray. *Religion, Brain, and Behavior* 1 (2011): 46-99.

Schultz, W. *Journal of Neural Transmission* (Vienna). doi: 10.1007/s00702-016-1510-0.2016.

Schwartenbeck, P., T.H.B. FitzGerald, and R. Dolan. *NeuroImage* 125 (2016): 578-586.

Schwartz, S., P. Maquet, and C. Frith. *Proceedings of the National Academy of Sciences USA*. 99 (2002): 17137-17142.

Seitz, R.J., M. Franz, and N.P. Azari. *Brain Research Reviews* 60 (2009): 368-378.

Shultz, S., and R.I.M. Dunbar. *Journal of Comparative Psychology* 124 (2010): 252-260.

Sidanius, J., F. Pratto, and C. van Laar. *Political Psychology* 25 (2004): 845-880.

Siegel, R.M., and E.M. Callaway. *PLOS Biol.* (2004). http://dx.doi.org/10.1371/journal.pbio.0020419.

Siever, L.J. *American Journal of Psychiatry* 165 (2008): 429-442.

Silk, J.B., S.F. Brosnan, J. Vonk, J. Henrich, D. J. Povinelli, A.S. Richardson, S.P. Lambeth, J. Mascaro, and S.J. Schapiro. *Nature* 437 (2005): 1357-1359.

Silk, J.B., and B.R. House. *Proceedings of the National Academy of Sciences USA* 108 (2011): 10910-10917.

Sirigu, A., and J.R. Duhamel. *Dialogues in Clinical Neurosci.* 18 (2016): 45-53.

Slavich, G.M., and S.W. Cole. *Clinical and Psychological Sciences* 1 (2013): 331-348.

Slon, V., C. Hopfe, C.L. Weiß, F. Mafessoni, M. de la Rasilla, et al. *Science* 27 Apr (2017). eaam9695. doi: 10.1126/science.aam9695.

Smith, K.L., K.C. Berridge, and J.W. Aldridge. *Proceedings of the National Academy of Sciences USA* 108 (2011): E255 – E264.

Steger, M.F., B.M. Hicks, R.F. Krueger, and T.J. Bouchard. *Journal of Positive Psychology* 6 (2011): 181-191.

Striedter, G.F. *Behavioral and Brain Sciences* 29 (2006): 1-36.

Stringer, J.M., S. Barrand, and P. Western. *Reproduction* 146 (2013): R37-R48.

Surbeck, M., C. Boesch, C. Girard-Buttoz, C. Crockford, G. Hohmann, and R.M. Wittig. *Annals of the New York Academy of Sciences* 907 (2000): 97-113.

Surbeck, M., C. Boesch, C. Girard-Buttoz, C. Crockford, G. Hohmann, and R.M. Wittig. *American Journal of Primatology* Jun; 79(6). (2017) doi: 10.1002/ajp.22641.

Szechtman, H., and E. Woody. *Psychological Reviews* 111 (2004): 111-127.

Sztajer, S. *Anglojęzyczny Suplement Przeglądu Religioznawczego* [*English-language Supplement of Religious Studies Review*] no.1, 2012.

Tattersall, I. *Collegium Antropologicum* [*College Anthropology*] 28, Suppl. 2 (2004): 1-8.

Tattersall, I. *Theory on Biosciences* 129 (2010): 193-201. doi: 10.1007/s12064-010-0093-9.

Tattersall, I. *Neuroscience and Biobehavioral Reviews* (2017): 1-8. http://dx.doi.org/10.1016/j.neubiorev.2017.01.043.

Teles, M.C., S.D. Cardoso, and R.F. Oliveira. *Frontiers in Behavioral Neurosciences* 10:16 (2016). doi:10.3389/fnbeh.2016.00016.

Thalbourne, M.A., and P.S. Delin. *Parapsychology* 58 (1994): 3-38.

Tognini, P. *Frontiers in Cellular Neuroscience* 11 (2017): 1-8.

Tomasello, M. *The Cultural Origins of Human Cognition.* Harvard University Press, 1999.

Tovote, P., M.S. Esposito, P. Botta, F. Chaudun, J.P. Fadok, M. Markovic, S.B.E. Wolff, C. Ramakrishnan, L. Fenno, and K. Deisseroth. *Nature* 534 (2016): 206-212.

Tremaroli, V., and F. Bäckhed. *Nature* 489 (2012): 242-249.

Triarhou, L.C., and M. del Cerro. *European Neurology* 56 (2006): 189-198. doi: 10.1159/000096425.

Triarhou, L.C. *Neuroscience and Biobehavioral Reviews* 32 (2008): 984-1000. doi: 10.1016/j.neubiorev.2008.03.013.

Tsankova, N., W. Renthal, A. Kumar, and J. Nestler. *Nature Review in Neuroscience* 8 (2007): 355-367.

Tunga, J., L.B. Barreiro, Z.P. Johnson, K.D. Hansen, V. Michopoulos, D. Toufexis, K. Michelini, M.E. Wilson, and Y. Gilada. *PNAS* 109 (2012): 6490-6495. www.pnas.org/cgi/doi/10.1073/pnas.1202734109.

Turnbaugh, P.J., R.E. Ley, M. Hamady, C. Fraser-Liggett, R. Knight, and J.I. Gordon. *Nature* 449 (2007): 804-810. doi:10.1038/nature06244.

Turner, AM, and W.T. Greenough. *Brain Research* 329 (1985): 195-203.

Vance, G.T., H.H. Maes, and K.S. Kendler. *Journal of Nervous and Mental Diseases* 198 (2010): 755-761. doi: 10.1097/NMD.0b013e3181f4a07c.

Vance, G.T. *Encyclopedia of Psychology and Religion* Springer, 1 (2014): 704-705. (Ed. Leeming D.E. doi: 10.1007/978-1-4614-6086-2_260.

van Elk, M., B.R. Rutjens, J. van der Pligt, and F. van Harreveld. *Journal of Religion Brain & Behavior*. 6 (2016): 4-33.

van Schaik, C.P. *The Primate Origins of Human Nature*. Hoboken, N. Jersey, USA: John Wiley & Sons, Inc, 2016.

van Schaik, C.P., K. Isler, and J.M. Burkart. *Trends in Cognitive Sciences* 16 (2012): 277-284.

van Schaik, C.P., J. Burkart J., L. Damerius, S.I. Forss, K. Koops M.A. van Noordwijk, and C. Schuppli. *Philosophical Transactions of the Royal Society* (B) (2016). doi:10.1098/rstb.2015.0183.

Varki, A., and T.K. Altheide. *Genome* 15 (2005): 1746-1758.

Varki, A., and D.L. Nelson. *Annual Review on Anthropology* 36 (2007): 191-209.

Varki, A., D.H. Geschwind, and E.E. Eichler. *Nature Reviews Genetics* 9 (2008): 749-763. doi:10.1038/nrg2428.

Veblen, T. *The Theory of the Leisure Class*. 1899, http://socserv2.mcmaster.ca/~econ/ugcm/3ll3/veblen/leisure/index.html.

Volkmar, F.R, and W.T. Greenough. *Science* 176 (1972): 1445-1447.

Vygotsky, L.S. *El Desarrollo de los Procesos Psicológicos Superiores* [*The Development of Higher Psychological Processes*], Ed. Critica, Barcelona, Spain, 1978.

Wadhwa, V. https://www.marketwatch.com/story/with-genetically-edited-babies-a-scientist-transgresses-a-moral-boundary-2018-11-26.

Waller, N. G., B.A. Kojetin, T.J. Bouchard, D.T. Lykken, and A. Telelgen. *Psychological Science* 1 (1990): 138-142.

Weaver, L.T., M.F. Laker, and R. Nelson. *Archives of Disease in Childhood* 59 (1984): 236-241.

Weaver, I.C., N. Cervoni, F.A. Champagne, A.C. D'Alessio, S. Sharma, J.R. Seckl, S. Dymov, M. Szyf, and M.J. Meaney. *Nature Neuroscience* 7 (2004): 847-854.

Weaver, I.C., M.J. Meaney, and M. Szyf. *Proceedings of the National Academy of Sciences USA* mc549A.103 (2006): 3480-3485.

Weaver, J. *PLOS Biology* 12: (2014): 1-2. e1002001.

Weizman, E. (2016) http://www.e-flux.com/architecture/superhumanity/68645/are-they-human/.

Westlye, L.T., T. Kaufmann, D. Alnæs, I.R. Hullstein, and A. Bjørnebekk. *Neuroimage Clinical* 13 (2017): 62-69.

Wheatley, T., O. Kang, C. Parkinson, and C.E. Looser. *Social and Personality Psychology Compass* 6/8 (2012): 589-606. 10.1111/j.1751-9004.2012.00450.x.

Wheeler, B.C., B. Tiddi, and M. Heistermann. *Animal Behavior* 93 (2014): 49-58.

Whitehouse, H. *Grounding the Social Sciences in the Cognitive Sciences* (R. Sun, ed.). Cambridge, Mass.: MIT Press, 2012.

Whitehouse, H., P. Francois, P.E. Savage, T.E. Currie, K.C. Feeney et al. *Nature* 2019.

Whiten, A. *Annual Reviews in Psychology* 68 (2017): 129-154.

Whiten, A. *Interface Focus* 7 (2017): 20160142. http://dx.doi.org/10.1098/rsfs.2016.0142. doi: 10.1016/j.neubiorev.2016.12.018.

Whiten, A., J. Goodall, W.C. McGrew, T. Nishida, V. Reynolds, Y. Sugiyama, C.E. Tutin, R.W. Wrangham, and C. Boesch. *Nature* 399 (1999): 682-685.

Whiten, A., and C.P. van Schaik. *Philosophical Transactions of the Royal Society* B 362 (2007): 603-620.

Whiten, A., and E. van de Waal. *Neuroscience and Biobehavioral Reviews* Dec 26 (2016): S0149-7634(16)30250-0.

Wightman, G.J. https://reader.paperc.com/books/the-origins-of-religion-in-the-paleolithic/800619/cover; Rowman and Littlefield, 2015, London, England.

Willyard, A.K., and A. Norenzayan. *Cognition* 129 (2013): 379-391.

Wilson, D.D., D. Near D., and R.R. Miller. *Machiavellian Intelligence II: Extensions and Evaluations*. Whiten A., Byrne R.W. (eds.), Cambridge University Press, 1996.

Wiltermuth, S.S., and C. Heath. *Association of Psychological Science* 20 (2009): 1-5.

Willyard, C. *Nature* 542 (2017): 406-408. doi:10.1038/nrmicro1978.

Wismer Fries, A.B., T.E. Ziegler, J.R. Kurian, S. Jacoris, and S.D. Pollak. *PNAS* 102 (2005): 17237-17240.

Woese, C. *Proceedings of the National Academy of Sciences* USA 95 (1998): 6854-6859.

Wolf, Y.I., and E.V. Koonin. *Bioessays* 35 (2013): 829-837.

Woody, E.Z., V. Lewis, L. Snider, H. Grant, M. Kamath, and H. Szechtman. *Behavioral and Brain Functions* 1 (2005): 11. doi:10. 1186/1744-9081-1-11.

Wrangham, R.W. *American Journal of Physiology and Anthropology Suppl*. 29 (1999): 1-30.

Wrangham, R.W., and L. Glowacki. *Human Nature* 23 (2012): 5-29. doi:10.1007/s12110-012-9132-1.

Wu, T.P., T. Wang, M.G. Seetin, Y. Lai, S. Zhu, K. Lin, Y. Liu, S.D. Byrum, S.G. Mackintosh, M. Zhong et al. *Nature* 532 (2016): 329-333. doi:10.1038/nature17640.

Xygalatas, D., P. Mitkidis, R. Fischer, P. Reddish, J. Skewes, A.W. Geertz, A. Roepstorff, and J. Bulbulia. *Psychological Science* 24 (2013): 1602-1605.

Zilles, K. "Evolution of the Human Brain and Comparative Cyto- and Receptor Architecture". In: *From Monkey Brain to Human Brain*. (Eds. Dehaene S., Duhamel J.R., Hauser M.D., Rizzolatti G.), Cambridge, MA, MIT Press, (2005): 41-56.

Zilles, K., and N. Palmero-Gallagher. *Frontiers in Neuroanatomy* 11 (2017): 78. doi: 10.3389/fnana.2017.000

About the Author

Jorge A. Colombo, MD, PhD

Dr. Colombo obtained his MD and PhD at the University of Buenos Aires, Argentina. He completed his postgraduate studies at the Brain Research Institute, University of California at Los Angeles (UCLA) in the United States. He is a retired Principal Investigator of Neuroscience at the National Research Council of Argentina (CONICET) and a former Associate Professor and tenured Full Professor at the College of Medicine at the University of South Florida (USF) in the USA.

Dr. Colombo is a former fellow at the Ford Foundation (USA) and the Foundation´s Fund for Research in Psychiatry (USA). He is also a former visiting professor of the National Institute of Health (USA) while researching in Sweden, the Alexander von Humboldt Foundation (Germany), the British Royal Society (England), the Deutscher Akademischer Austauschdienst (DAAD Germany), and the Iberdrola (Spain). Dr. Colombo served as Acting Full Professor of Neurophysiology at the School of Psychology at the University of Buenos Aires, Argentina, as well as Full Professor of Psychophysiopathology at the University of Belgrano, Argentina.

Dr. Colombo is the Founder and Director Emeritus of the Unit of Applied Neurobiology (UNA) (Centro de Educación Medica e

Investigaciones Clinicas-CONICET). He is the President and Scientific Director of Fundacion Conectar (for the Neurosciences). He has authored numerous scientific publications in books and international journals.

INDEX

A

B

C

D

E

F

T

W